NORTH AMERICAN WILDLIFE

STANLEY KLEIN

NORTH AMERICAN WILDLIFE

STANLEY KLEIN

GALLERY BOOKS
An Imprint of W. H. Smith Publishers Inc.
112 Madison Avenue
New York City 10016
A Bison Book

Published by Gallery Books
An imprint of W. H. Smith Publishers Inc.
112 Madison Avenue
New York, New York 10016

First Published in 1983
This abridged edition produced in 1984 by
Bison Books Corp.
17 Sherwood Place
Greenwich, CT 06830

Printed in Hong Kong

1 2 3 4 5 6 7 8 9 10

ISBN 0-8317-6437-6

Page 1; **A playful though defiant yearling Raccoon (*Procyon lotor*) is captured by our camera as he prowls by night in the heart of the city of San Francisco.**

Page 2-3: **A six-point Whitetail buck (*Odocoileus virginianus*) looks up from his grazing in a meadow in Montana's Glacier National Park.**

Above: **A herd of White-tailed Deer (*Odocoileus virginianus*) perk up their ears at the sound of a possible predator in the rugged foothills east of the Rocky Mountains**

CONTENTS

Natural Features of North America
Showing Major Mountains and Larger Canadian and American National Parks and Monuments
ASIA
BERING SEA
ALEUTIAN ISLANDS
ARCTIC OCEAN
GREENLAND ICE SHEET
Queen Elizabeth Islands
Victoria Island
Baffin Island
Bering Land Bridge NM
Noatak NM
Gates of the Arctic NM
Brooks Range
Yukon River
Denali NM/Mt. McKinley NP
Katmai NM
Lake Clark NM
Mt. McKinley
Anchorage
Yukon-Charley NM
Wrangell-St. Elias NM
GULF OF ALASKA
Mt. Logan
Glacier Bay NM
Mac Kenzie R
Great Bear Lake
Great Slave Lake
ROCKY MOUNTAINS
Fraser R
L. Athabasca
Wood Buffalo NP
HUDSON BAY
LABRADOR SEA
NORTH PACIFIC OCEAN
Vancouver
Puget Sound
Olympic NP
Mt. Rainier
Jasper NP
Banff NP
Yoho NP
Kootenay NP
Waterton Lakes NP
Glacier NP
Prince Albert NP
L. Winnepeg
Riding Mt NP
Columbia R
Flathead L
Ft Peck Res
Cascade Range
Yellowstone NP
Yellowstone R
Missouri River
Cape Mendocino
Mt. Shasta
Sierra Nevada
GREAT BASIN
Great Salt Lake
San Francisco Bay
San Francisco
Yosemite NP
Rocky Mtn NP
Platte R
Sequoia/Kings Canyon NP
Mt. Whitney
Death Valley NM
Canyonlands NP
Grand Canyon NP
Colorado River
Arkansas River
L. Superior
L. Michigan
L. Huron
L. Ontario
L. Erie
St. Lawrence River
Montreal
Gulf of St Lawrence
Chicago
Cape Cod
New York
Mississippi River
Ohio R
Great Smoky Mtns NP
APPALACHIAN MTNS
Chesapeake Bay
Tennessee River
Red River
Cape Hatteras
NORTH ATLANTIC OCEAN
Big Bend NP
Rio Grande
GULF OF MEXICO
L. Okeechobee
Everglades NP
CENTRAL AMERICA
CARIBBEAN SEA
Ice
Tundra
Mountainous
Northern Coniferous Forest
Mixed Coniferous & Deciduous Forest
Plains & Grasslands
Deserts & Chaparrel

INTRODUCTION

With only about 5000 species, mammals are not the most numerous animals. They are, however, the most dominant animals on this planet. Man, the most dominant animal, is a mammal. The first mammals are believed to have evolved about 140 million years ago. However, for more than 100 million years, mammals were eclipsed by reptiles, particularly the large dinosaurs. Some 70 million years ago, the dinosaurs became extinct. The mammals had the adaptations that allowed them to survive in the conditions that caused the demise of the dinosaurs and to become the dominant life forms. Today, mammals are found in every habitat and climate from polar regions to tropics. Mammals live in every environment—oceans, bodies of fresh water, mountains, plains, forests, deserts and more. There are mammals that swim, fly and burrow in the ground.

Mammals, like birds, are able to maintain a relatively consistent body temperature regardless of the outside temperature. Unlike reptiles which become inactive in cold conditions, most mammals can remain active in a wide range of temperatures. The minimum temperature at which the animal can survive varies from one species to another, and some mammals survive cold winter conditions by going into a state of suspended activity called hibernation.

Extremely important to the success of the mammals is the way they bear their young and take care of them. Mammals develop inside the female in a specialized structure called the uterus. The developing young are less likely to be consumed by predators than are eggs. A pregnant mammal has a chance to get away from a predator and, if she does, she saves not only herself but also her unborn offspring. A bird or reptile cannot carry eggs away from a predator. It can only try to fight off the predator or hide the eggs when they are laid. Mammals produce milk for their young in special glands. The ability to produce milk is a singular characteristic of mammals. The word mammal comes from the Latin *mamma* which means breast. Not all mammals have breasts, but almost all have teats or nipples on which the young suck to obtain milk in the early stages of life.

Mammals have more highly developed brains than either birds or reptiles. Some mammals are capable of highly sophisticated activities such as coordinating the hunting of prey. Lions and wolves, for example, hunt in groups and exercise what seems to be highly developed strategy. Some animals such as chimpanzees and dolphins may be able to reason and solve problems

Mammals take more care of their young than other forms of animal life. (Birds also care for their young, but practically all avian parental behavior is mechanical action triggered by instincts.) Much mammalian parental behavior is also instinctive, but many mammals plan their actions, exercise choices and teach their young survival skills at a higher level.

The time mammals take care of their young varies from a few days in some rodents to decades in humans. The care given by mammals makes up for their relatively slow rates of reproduction. A mouse that has five or six litters of 10 to 15 young a year is considered to be a prodigious producer. Most mammals have much longer gestation periods and have only a few young at a time. A single fish can lay hundreds of thousands of eggs at a time. However, only a very small fraction of the fish eggs will survive to become adult fish. Young mammals, on the other hand, have a much better chance for survival because of parental care.

In general, the larger the mammal, the longer the gestation period and the smaller the number of offspring. The gestation period of the elephant, for example, is almost two years, and one baby elephant is the usual result. Mice have gestation periods measured in weeks and produce litters containing many offspring.

In addition to warm-bloodedness and the production of milk, mammals have:

- bodies with hair or fur
- four-chambered hearts
- lungs used for breathing
- seven cervical (neck) vertebrae (some exceptions)
- highly developed nervous systems
- young which are born alive (except monotremes)

Mammals can be divided into three groups on the basis of how they reproduce. The monotremes are those mammals that lay eggs, such as the Platypus. The marsupials are the pouched mammals. They give birth to very immature young that must find their way to the pouch where their development is completed. The only North American marsupial is the opossom. Other familiar marsupials are Kangaroos and other Australian mammals such as Koalas and Wombats. Most mammals are placental. Placental mammals develop in the uterus and receive their nourishment through a structure called the placenta. Although the young are born helpless, they are in a much higher state of development than is the case with marsupials.

ANTELOPE, AMERICAN: see Antelope, Pronghorn

ANTELOPE, PRONGHORN: *(Antilocarpa americanus)*

The Pronghorn lives in open grasslands and bunch-grass and sage-brush areas of the western United States. Its range includes Wyoming, Colorado, Utah, Arizona, New Mexico, Nevada, western Texas, eastern Montana, southeastern Oregon and small areas in Washington and northern California. It is also found in southern California and the extreme southern portion of Idaho, northern Mexico along the Texas border and Baja California.

A medium-sized deer, the Pronghorn is about 3 to 3.5 feet (.9 to 1 m) high at the shoulder, and from 4 to 4¾ feet (1.2 to 1.4 m) long. Males weigh from 80 to 160 pounds (36 to 170 kg) and females weigh from 70 to 110 pounds (32 to 50 kg). The upper and outer parts of the body are pale tan to reddish-tan, while the underside, rump patch, cheeks and lower jaw are white. Two white blazes run across the throat. The bucks have a distinctive broad, black band that runs from the eyes, down the snout to the nose, and a black crescent across the neck and black antlers. One foot to 1⅔ feet (.3 to .5 m) long at maturity, the antlers curve back and slightly inward. On each side there is a broad, short prong that juts forward and slightly upward. Does have short antlers no more than 3 or 4 inches (7.6 to 10 cm) long that usually lack a prong.

The Pronghorn is well adapted for its life in open grasslands. It can run at speeds of 70 mph (112 kph) for bursts of three or four minutes, a pace that makes it the fastest animal in North America and one of the fastest in the world. It moves along easily at speeds of 30 mph (48 kph), and sustained runs at 45 mph (77 kph) are not unusual. Its large eyes have a wide field of vision, and it detects motion at distances of four miles (6 km) or more. It is also a strong swimmer. When it is alarmed, the rump hairs become erect, forming a 'flash' that can be seen for great distances and probably serves as a danger signal.

***Above:* A Pronghorn or American Antelope buck, the fastest mammal in North America.**

The summer graze includes a variety of grasses and cacti. In winter it browses mostly on sagebrush, but will take other plants if no sagebrush is available.

The sexes form separate, small herds in the summer. Does and fawns herd separately from groups of bucks, yearlings and two-year-old males. During the breeding season, which begins in the autumn, older males compete to form harems of about 20 does which are maintained in territories. Bucks defend their territories with snorts, chases, postures and occasional fights. The young are born in May or June. A doe's first breeding usually results in one fawn; subsequent births are usually twins and sometimes triplets. If the doe has more than one fawn, she delivers them at spots several hundred yards apart. The doe hides the fawns in high grass or brush, and grazes away from them to distract predators.

In the 1920s the Pronghorn almost became extinct, mainly because of fencing put up by ranchers which restricted their movements. Cattlemen also killed large numbers of them in the belief they reduced graze for cattle. Management programs, including transporting animals to new areas, have increased their number to almost 500,000.

Pronghorn are similar to the White-tailed Deer (*Odocoileus virginianus*), which do not have the prong on their antlers.

ARMADILLO, NINE-BANDED *(Dasypus novemcinctus)*

The Nine-banded Armadillo's range includes Texas, Oklahoma, Arkansas, Louisiana, Mississippi, southern Alabama, southern Georgia and Florida. The armadillo was named by early Spanish explorers of the Americas who called it 'little armored thing' because of the bony plates in its upper body skin. Unlike its prehistoric forebears, which could be as large as a small elephant, the armadillo today is about 2 feet (61 cm) long, including its tail, and weighs about 15 pounds (7 kg).

Above: **The Badger is known for the white badge-like marking on its forehead.**

The Nine-banded Armadillo, common to the United States and South America, is so called because the hinges of its shell consist of nine narrow bands of armor which move upon one another to give the animal some flexibility.

Armadillos have long, narrow tongues which they use to lick up insects, earthworms and land snails, their principal food. They are also known to eat small snakes and toads.

Above: **A Nine-banded Armadillo.**

Their powerful claws enable them to dig rapidly and they resort to tunnels and burrows for protection. Their teeth are small and rootless, set well back in the mouth, and thus ill-adapted to self-defense.

The Nine-banded Armadillo generally sleeps during the day and comes out at night to feed, moving quickly on land and swimming strongly when necessary. Its mating season is in July and August, followed by a comparatively long gestation period of nine months. There are usually four young to a litter, and they are invariably of the same sex and bear identical sets of genes. This fact has made them valuable to scientific researchers, a value augmented in 1971 when they were discovered to be one of the few creatures besides man known to contract leprosy. The Nine-banded Armadillo is well-developed and mobile at birth, but its skin is soft and hardens only gradually into the protective carapace.

Armadillo meat is considered a delicacy by the Indians of Mexico; in the United States it is eaten most frequently in Texas, where the animal is both popular and plentiful.

BADGER
(Taxidea taxus)

The badger is found in the United States in all states west of the Mississippi River except Alaska, Louisiana, Arkansas and parts of Missouri, Texas, California, Oregon and Washington. The range extends southward into Mexico. East of the Mississippi River, its range includes Wisconsin, Illinois, Indiana, Michigan and most of Ohio. In Canada it is found in eastern British Columbia, most of Alberta, southern Saskatchewan and Manitoba and in extreme southern Ontario.

The badger is from 1.7 to 2.9 feet (.5 to .9 m) long and weighs from 8 to 25 pounds (3.5 to 11 kg). Its flat body is wider than it is high, and it has short legs that bow outward. A lightly upturned snout gives it somewhat of an inquisitive appearance. Its shaggy fur is a grizzled gray to brown, and it has a white stripe running between the shoulders. The fur on the short bushy tail is yellowish.

The badger is mostly nocturnal in areas near human populations, but it can sometimes be seen in the daytime in open plains, farmlands and the edges of woodlands, moving in a shuffling, rocking-from-side-to-side gait. It can move at a fair, but clumsy, trot, but it does not depend on speed to get its food or to escape from potential predators. Feeding primarily on ground squirrels, gophers, rats and mice, it takes these animals by digging out their burrows. The intended prey sometimes escapes by scampering out another exit, a tactic the badger may attempt to circumvent by burrowing into an empty burrow and waiting for the tenant to return. Badgers sometimes lose potential catches to coyotes who wait at an exit hole for the prey to be chased out of the burrow by the badger.

Badgers also take food on the surface, including birds, various invertebrates and rattlesnakes, which appear to be a particular favorite. The badger was long believed to be immune to the rattlesnake's venom, unless it was bitten on the nose but in fact, its thick fur prevents the snake's fangs from penetrating its skin anywhere else. It also eats carrion, and caches excess kills by burying them. The badger is an excellent swimmer, but it seems to take to the water more for comfort than in pursuit of prey. On particularly hot days, it may lie down in shallow water.

The badger is a fastidiously clean animal. It buries its droppings, and when it is at rest, cleans and grooms itself almost constantly with its tongue, swallowing the loose hair. It digs out a den or burrow with an elliptical entrance that accommodates the animal's shape. During very cold weather the badger holes up in a chamber deep inside the burrow and goes into a deep sleep, but does not hibernate.

Mating takes place in late summer. Litters of two to five young are born, after delayed implantation, in March and April. The young leave to go out on their own when the mother is ready to mate again.

The badger is seldom the prey of other meat-eaters. Thick fur, heavy neck muscles, and a loose, tough hide enable it to withstand biting and clawing attacks. It fights back with tooth, claw and a foul-smelling musky secretion, but it prefers to retreat if possible by backing into a burrow, holding off the adversary for only as long as it takes to plug up the burrow. If there is no burrow around, it may start digging one at a remarkable rate of speed, throwing dirt in the face of its enemy.

Badger hair is still considered to be superior to synthetic fibers for producing the best quality paint brushes. The coarse hairs were once used to make shaving brushes, a use that has declined with the advent of aerosol shaving cream cans. Badgers seldom take a domestic animal, but they are considered to be a nuisance by cattlemen, because cattle, horses and sometimes the cattlemen themselves fall into their burrows and break legs. Farmers, however, consider them to be beneficial animals because of the large numbers of rodents they destroy. Badger hunting was once a popular sport, and a breed of dog, the low-slung, short-legged Dachshund – which means 'badger hound' in German – was developed for the purpose of going into burrows after the badger. The word 'to badger,' which means to torment or harass, stems from badger hunting.

There is no other North American mammal quite like the badger. The face of the Coati (*Nasua nasua*) has a vague resemblance to that of the badger, but is not really similar.

Above both: **A Common Brown Bat (*Eptesicus fuscus bernardinus*) male.**

BAT, BIG BROWN
(Eptesicus fuscus)

The Big Brown Bat is found in all of the United States except the southern tip of Florida and south central Texas. It is found in southern Canada from coast to coast.

Somewhat larger than the Myotis Bats, the Big Brown Bat is from 4⅛ to 5 inches (10.5 to 13 cm) long and weighs about ⅝ ounce (18 g). Those that live in desert areas have light brown coloration, while those in forests tend to be a glossy brown. The undersides are a pale tan color, and the interfemoral membrane is black.

The Big Brown Bat roosts in small groups of five or six in a variety of sheltered spots including buildings, caves, mines, storm sewers and hollow trees.

The Big Brown Bat feeds mostly on large beetles, but it also takes flies, moths, wasps, leafhoppers and other kinds of insects. Thought to be the fastest flyer among the bats, it has been clocked at 40 mph (64 kph).

One or two young are born in May or June. Over the winter they hibernate, making use of body fat stores.

The Evening Bat (*Nycticeius humeralis*) is a similar species.

BAT, CALIFORNIA LEAF-NOSED; LEAF-NOSED BAT
(Macrotus californicus)

In the United States, the California Leaf-nosed Bat is limited to southern Arizona, southern California and the extreme southern tip of Nevada. It is the only leaf-nosed bat found in the United States.

The California Leaf-nosed Bat has grayish to dark brown fur above and white or nearly-white fur below. It is from 3⅜ to 3⅝ inches (8.5-9 cm) and weighs from ⅜ to ½ ounces (9-14 g). Like all leaf-nosed bats, it has an erect flap of tissue on its nose that is thought to function in the emission of ultrasonic sounds for echolocation.

The California Leaf-nosed Bat rests

by day in colonies of about 100 individuals. Frequently used daytime roosting areas are abandoned mine tunnels, close to the entrances. It emerges at night, hunts for about an hour until its stomach is full, and then retires to a night roost in some sheltered area where it stays until it returns to the daytime roost.

The California Leaf-nosed Bat eats flying and ground insects. It is often seen to hover over the ground and then swoop down to pick up an insect it has detected through echolocation.

The sexes generally stay apart, except at mating time. Young are born in May or June.

Similar North American species include the Hognosed Bat which has its flap over a rather long, slender nose, and the Longnosed Bat which, as its name suggests, has an even longer nose than that of the Hognosed Bat.

BAT, HAIRY-LEGGED VAMPIRE; VAMPIRE BAT *(Diphylla ecaudata)*

The Vampire Bat's normal range is from Mexico south through Central America and South America. Rarely found in the United States, the very few documented sightings have been in Texas. All species of Vampire Bats are found only in the western hemisphere.

The Hairy-legged Vampire Bat is from 2⅝ to 3⅜ inches (6.6 to 8.6 cm) long. Its fur is reddish-brown to dark brown above and lighter below. The middle upper incisor teeth are modified into large, sharp piercing tools. Since this bat lacks a tail, the interfemoral membrane has more fur on it than that of most other bats.

The Hairy-legged Vampire Bat's diet is limited to the blood of other warm-blooded animals. By day it roosts in caves, mines and other secluded areas in small groups, usually of no more than three, sometimes a dozen or so. At night it seeks out animals such as chickens, turkeys, ducks, geese, cattle, horses and pigs. Unlike most bats, the Hairy-legged Vampire Bat has large eyes and well-

***Above:* Both Red Bat males and females can be distinguished by their color.**

developed vision, and a rather poor echolocation sense.

One of the few bats that can use its wings for walking, it lands near its prey and walks over to it in a scuttling motion, using its wings. If the prey is small, such as a chicken, it will climb up the leg. It might jump onto a larger animal such as a cow. It is light enough so that it seldom alerts the animal as it moves about on its surface. With its specialized incisor teeth, it makes a wound that is usually too superficial for the victim to notice. The bat then laps up the blood that flows freely from the wound because of an anticoagulant in the bat's saliva.

BAT, LITTLE BROWN: see Myotis, Little Brown

BAT, RED *(Lasiurus borealis)*

The Red Bat is found in all of the United States except southern Florida and the Rocky Mountain areas. It is also found in most of Canada except the northern and Pacific Coast regions.

A richly colored bat, the Red Bat is one of the few in which there is a difference in coloration between the sexes. The fur of the male is a bright red or orange-red, while that of the female is a dull red. The fur of both sexes is a frosty white on the back and underside. The Red Bat is from 3¾ to 5 inches (9.5 to 13 cm) long and weighs about ½ ounce (14 g).

The Red Bat roosts in dense foliage rather than in caves or other sheltered areas, usually hanging from 4 to 10 feet (1.2 to 3 m) above the ground. The roosting sites are chosen carefully to provide shade from above and the sides and an opening for a take-off below.

In the evening they leave their roosts to hunt for insects. Fast, steady flyers, they take a variety of insects, including ants and beetles as well as flying insects such as moths and flies. Red Bats appear to have definite hunting routes which they work over repeatedly. Those in the northern parts of the range migrate south in winter.

The females bear two to four young in June. Females with young roost some 10 to 20 feet (3 to 6 m) from the ground. The young cling to the mother until they become so heavy that they prevent the mother from taking off.

Similar species include the Yellow Bat, Hoary Bat (*Lasiurus cinereus*), and Seminole Bat (*Lasiurus seminolus*).

BAT, SEMINOLE: see Bat, Red

BEAR, ALASKAN BROWN: see Bear, Grizzly

BEAR, BLACK
(Ursus americanus)

Above: **The Black Bear has a clumsy walk but is able to run at great speed.**

Bears are among the chief attractions at national parks such as Yellowstone; but it has become more difficult to observe them under recent park policies aimed at visitor safety. Although the bear most likely to be seen by the tourists visiting the park is the Black Bear, chances are that the color of the bears they see is more of a brown than black. The Black Bear, also called the Cinnamon Bear, exists in several color phases. In eastern North America, the animals are black with a brownish tinge; in the west they are cinnamon or brown in color, with a white blaze on the chest. A 'blue' or 'glacier' phase is found around Yukatat Bay in southwestern Alaska, and those found on Gribble Island in Alaska are almost white. The face is brown in all the color phases.

The smallest of the North American bears, when on all fours, the Black Bear is from 5 feet 6 inches long (1.5-1.8 m), some 2 feet 3 inches high (.6-.9 m) at the shoulders and weighs from 200-600 pounds (90-270 kg).

The Black Bear is found in most of Canada and Alaska. In the eastern United States it is found in northern New England, eastern New York, central Pennsylvania and in the Allegheny Mountains, northern Minnesota, Michigan and Wisconsin. It also occurs in Florida, Louisiana, Arkansas, southern Missouri and Illinois. In the west it is found in the Rocky Mountain States, western Washington and Oregon, extreme northern California and in the Sierra Nevada mountain range. Its range extends into the Mexican state of Chihuahua. Although it is quite abundant in mountainous territory, it is seldom seen at heights higher than around 3000 feet (900 m).

The Black Bear is usually nocturnal, but it can often be seen at midday, especially in areas where they have learned to expect handouts from people. Although classified as a meat-eater, bears will eat almost anything. Their diet includes berries and other fruit, roots, insects and their grubs, nuts, eggs, small animals and carrion. In the spring, they peel the bark off trees to get at the green, growing area under the bark. They tear apart rotting logs to get at insect grubs, beetles and other insects. They also rip open bee trees to get at the comb, honey and bee larvae. In areas near human habitation, they are commonly seen in garbage dumps. Good fishermen, they often wade into streams and rivers to catch fish with their paws.

Bears are excellent swimmers and tree climbers. Usually, their method of moving is a clumsy-looking shuffle. They are, however, capable of running as fast as 30 mph (48 kph). They retire to a sheltered den during the winter. Although they spend much of the winter sleeping, their winter retirement is not a true hibernation. They wake up from time to time, while a hibernating animal will not revive even if it is handled.

Black Bears are usually solitary animals except at mating time, which occurs in June and July. Females usually bear a litter every other year. The cubs are born in the winter in January and February. Most often two cubs are born; triplets are born occasionally. A first-time mother usually gives birth to one cub which weighs about ½ pound (.2 kg) when it is born. Emerging from the den with its mother, it will stay with her for about a year. Females are ready to breed when they are about three years old.

Black Bears tend to keep to themselves and are usually not a danger to people. Females with cubs, however, have a strong protective instinct and

Above: A Grizzly Bear and her cubs.

can be quite dangerous. Bears that live in areas around human habitation lose their fear of humans and can be dangerous. Attacks have occurred around garbage dumps and in national parks where people try to feed the Black Bears. They sometimes raid campgrounds, looking for food. Occasionally, people are killed or seriously injured. Bear hunting is a popular sport in some areas. The meat is considered to be excellent, although it must be thoroughly cooked to avoid getting a parasitic infection. The hides are used for rugs and ornamental fur pieces.

BEAR, BROWN: see Bear, Grizzly

BEAR, CINNAMON: see Bear, Black

BEAR, GRIZZLY; KODIAK BEAR, BROWN BEAR *(Ursus arctos)*

American folklore is filled with stories and songs of heroic characters like Davy Crockett who wrestled and killed ferocious, 20-foot-tall Grizzly Bears with a hunting knife. While the Grizzly Bear is indeed the largest of American bears and can be dangerous, the stories and legends about the size and ferocity of this animal tend to be exaggerated.

When Europeans first came to North America, the range of the Grizzly Bear included the entire western half of the United States and Canada. Today, this bear is limited to wilderness areas in Alaska and in Canada: in British Columbia, the Yukon Territory, western Alberta and the Northwest Territories. Rare in the United States today, the Grizzly Bear is limited to high mountain areas in northern Idaho, eastern Montana and the higher ranges of the Rocky Mountains in Wyoming and Colorado. Most of them are in national parks such as Yellowstone and Glacier. The Alaskan Brown and Kodiak varieties are found on the Alaskan coast and on islands along the coast. The name of the Kodiak Bear comes from Kodiak Island, Alaska.

Grizzly Bears, while on all fours, are from 6 to 7 feet (1.8-2.1 m) long, about 4¼ to 4½ feet (1.2-1.3 m) high at the shoulders, and weigh from 300 to 1500 lbs (136-680 kg). The claws on the front paws are an impressive 4 inches (10 cm) long. The Kodiak Bear (Brown variety), or Alaskan Bear, reaches lengths of around eight feet (2.4 m) and weighs as much as 1700 lbs (770 kg). These bears are the largest of the terrestrial carnivores. Grizzly Bears are yellowish-brown to black. Often, the tips of the hairs are white, giving the bear its characteristic grizzled appearance. The profile of the face is somewhat concave and there is a definite hump above the shoulders. The Grizzly Bear walks in a low, clumsy stance, swinging its head back and forth as it moves. It can, however, move at speeds of 15-20 miles an hour (24-32 kph). The cubs can climb trees, but this ability is lost in adulthood.

Grizzly Bears are active day or night, roaming by themselves or in small family groups. Although classified as carnivores, they will eat almost anything they can get, including plants, large and small mammals and carrion. They dig out small rodents from their burrows. If a Grizzly Bear finds carrion, it will store the carcass in a hiding place, returning to it from time to time to eat. Grizzlies are also quite adept at catching fish. They gather along rivers to catch salmon as these fish swim upstream to spawn. They snap the fish out of the water with their jaws or pin them under the water with their forepaws. The Bears engage in vicious fights over territory.

***Above:* Grizzly or Brown Bears.**

In the winter they hole up in a den. Their winter sleep, which is not a true hibernation, is characterized by frequent awakenings and they sometimes actually leave the den during the winter months. They tend to return to the same den every year.

Grizzly Bears mate in June and July and the cubs are born in January. Females are ready to mate when they are three years old, and they generally have a litter every other year. The litters range from one to four cubs; the usual number is two. The cubs, which weigh about a pound (.45 kg) when they are born, stay with their mother for a little more than a year.

Although the Grizzly Bear has a reputation for ferocity, it usually minds its own business and avoids contact with humans. However, its behavior is unpredictable, and it should be considered dangerous if ever encountered. Many of these bears were killed by ranchers and farmers who believed they destroyed livestock. Today, few of these animals are in areas where they are a threat to domestic animals. The Grizzly is considered to be a species threatened with extinction.

Biologists disagree on whether all the bears commonly called Grizzly, Kodiak and Brown are the same or separate species. Some biologists say there are as many as 70 separate species of 'Grizzly' Bears while others say there is only one. Still other biologists put the number of species at some number between one and seventy. All of these bears, however, are closely enough related in appearance and habits to be considered together.

BEAR, KODIAK: see Bear, Grizzly

POLAR BEAR
(Ursus maritimus)

The Polar Bear, king of the North, may weigh 1000 pounds and measure eight feet from nose to tail. The large individual is five feet high at the shoulders and may tower 12 feet above the ground when erect on its broad hind feet. These magnificent animals are unbelievably powerful and unpredictable, a combination that makes them dangerous, especially when they are caught by surprise or when a female with cubs is approached. Furthermore, the polar bear may show little fear of people where protected from shooting. In Churchill, Manitoba, which lies on a polar bear migration route, the bears become accustomed to roaming through the center of town, poking their snouts into open doors and feeding at the dumps.

Through most of their wide range, however, the Polar Bears belong to the wilderness. They live in the Arctic regions of five countries: Canada, the United States, the USSR, Denmark (Greenland), and Norway. Problems for the Polar Bear intensified when men with guns began invading the Arctic. Today those five nations have an agreement to help protect the bears; joint commissions study the bears and cooperate in setting regulations. Even with this protection, Polar Bear numbers total probably fewer than 20,000.

These sea-going bears are seal

Above and right: **The Polar Bear is active throughout the year, unlike other bears.**

hunters and their major prey is the ringed seal, which weighs, when grown, 100 pounds or more. The polar bear has two basic hunting techniques. One is to lie in ambush beside the seal's breathing hole. As the seal surfaces to breathe, which it must do frequently, a giant paw flashes out and crushes the seal's skull. The other method for hunting is to slip up on sleeping seals. When a seal rests on the ice, it must be constantly alert, sleeping only a few minutes at a time, then lifting its head to survey the scene. While the seal is looking, the stalking bear freezes in place and waits for the seal's head to drop again. The bear advances cautiously, taking advantage of ice ridges to help hide its movements. The perfectly camouflaged bear often, but not always, gets close enough to rush the seal before it can escape to the water.

Polar Bears spend large parts of their lives drifting on ice floes. Traveling with the currents, they may cover several hundred miles a year. Meanwhile, the female about to bear young is snugly hidden in a den. She has moved inland and chosen a steep bank where the snow will drift and cover her well. She hollows out a room, keeping a breathing hole open.

In the dead of winter she gives birth usually to twins, each weighing about a pound and a half. The tiny cubs are blind and helpless, but by the end of winter they are big enough to leave the den with their mother. For two and a half years they stay with her, learning to capture food. The female gives birth only every second year, which leaves sufficient time to care for her cubs until they can survive alone.

The giant bear is well outfitted for life in the Arctic. It wears a thick coat of long, white hair which covers even the soles of its feet. Beneath the skin lies a layer of fat, sometimes four inches thick, that serves as both insulation and a source of energy. The bear can lumber along at speeds of 35 miles an hour. It is also at home in the water, where it can swim at six miles an hour and may travel for a hundred miles or more.

The Polar Bear possesses a sense of smell so keen that Eskimos insist the bear smells odors from miles away.

In recent times, biologists have made extensive field studies of Polar Bears to learn more about the bears'

Left: **The Beaver.**

needs and how to protect them from modern hazards. By using small transmitters which emit signals that can be followed by airborne biologists, scientists have brought us new information on how the bears live and where their travels take them.

BEAVER
(Castor canadensis)

When Europeans came to North America, the Beaver's range extended over most of what is today the United States and Canada, including Alaska. Excessive trapping soon diminished its range considerably. However, in recent years, as the result of reduced trapping, it has returned to much of its original range. Today it is found in all of the United States and Canada, except most of Florida and parts of Utah, Arizona, Nevada and California, and the extreme northern parts of Canada and Alaska. In Mexico, it is limited to the northern areas along the Rio Grande.

One of the larger rodents, the Beaver is dark brown except for its horizontally flattened, paddle-like tail which is black and scaly. It is from about 3 feet to 3.8 feet (.9-1.2 m) long, and generally weighs from 45 to 60 pounds (20-27 kg). Some specimens as heavy as 100 pounds (45 kg) are on record. The Beaver has large incisor teeth that continue to grow throughout the animal's lifetime; otherwise they would be worn away by the Beaver's constant gnawing on wood. The tail makes a loud slapping noise when it hits the water. Loud enough to be heard at considerable distances, it is thought to be an alarm signal.

The Beaver is well adapted for life in the water. It has webbed hind feet that enable it to swim with great efficiency. The tail is used as a rudder, freeing the paws to hold objects. The Beaver, which spends a great deal of time underwater, can remain submerged for as long as 15 minutes at a time. The Beaver is protected underwater by valves that close off the ears and nostrils and by skin flaps that seal the mouth, except for the incisor teeth which are used for carrying branches. The eyes are protected by clear membranes. The Beaver waterproofs its fur with secretions from its anal glands applied to the fur with its hind feet which are equipped with nails that serve as a comb. Additional protection is provided by a layer of fat.

The Beaver remains active throughout the entire year. Mostly nocturnal, it is seldom observed in daylight hours. Beavers live near or in bodies of fresh water such as rivers, ponds and streams. Those that live around large rivers dig burrows along the riverbank, with entrances below the water level. Those that live in smaller streams, ponds and lakes build much more elaborate homes.

The Beaver's home-building activity, which involves the construction of dams and lodges with underwater entrances, is one of the most fascinating animal behaviors. The dams are built to raise the water level in the stream or pond so that the lodge entrance will be under water. Both dam and lodge are built of wood, generally with branches of poplar, aspen, willow, birch or maple. The building materials are also the Beaver's food.

Beavers fell trees by gnawing around the trunk, biting out deep chunks. Trees with trunk diameters from two inches to six inches are usually selected, although Beavers are capable of gnawing down much larger trees. Trees with trunk diameters of 30 inches or more are known to have been felled by Beavers. The Beaver can cut down a tree at speeds a lumberjack would envy. A tree four feet in diameter can be felled in five minutes.

Tree branches, stones and mud are the main construction material for the dams and lodges. The Beaver eats the bark and then uses the stripped branches for building the dam. The lodge, also built of branches, has a floor of wood chips, and is generally 3 to 5 feet (.9-1.5 m) wide with a ventilation hole at the top. Additional branches are stored by sticking them into the bottom mud at some point near the lodge. Thus the Beaver has a ready food supply in winter when the pond is frozen over. Beavers are constantly repairing their dams and adjusting them for changing conditions. For example, in flood conditions they construct temporary spillways along the sides of the dams.

Thought to form lifetime pairs, Beavers mate in late January and early February. The kits are born in late spring to early summer. Litters vary from one to eight kits with four or five being the usual number. The kits, which weigh about a pound (0.5 kg) at birth, are adept swimmers by the time they are a week old. In the water the kits are sometimes ferried on the mother's back. On land they may be carried on the mother's tail. The upper chamber of the lodge serves as living quarters, with the ventilation hole concealed by sticks, and the lower chamber, beneath the water level, is a food storage area. Since the underwater tunnels provide the only access to the lodge, the beaver is safe from almost every predator.

Around the middle of the 19th century, practically every gentleman in America and Europe was expected to own a tall 'stovepipe' hat. Almost all of these hats were made of beaver fur, a circumstance that almost led to the extinction of these remarkable rodents.

The Beaver reached the peak of its economic importance in the early to middle 19th century. The pelts were much in demand for coats and ornamental trim as well as for hats. Great fortunes were made from Beaver trapping. The Beaver has made a remarkable comeback and is considered to be a pest in some areas by farmers. Their dams sometimes flood cropland and roads, and prevent salmon swimming upstream to spawn. However, the dams retard erosion and provide a feeding and drinking area for other life forms.

Similar rodent species are the Muskrat (*Ondatra zibethica*) and the Nutria (*Myocastor coypus*), both of which are smaller than the beaver. The River Otter (*Lutra canadensis*) is similar in that it lives in water.

BISON, AMERICAN BUFFALO
(Bison bison)

Although the Bison or American Buffalo never received serious consideration as the official animal symbol of the United States, it has come to be regarded as the unofficial symbol of the country. To many people, the Bison evokes America much as the Kangaroo is evocative of Australia. It is estimated that, in the century before Europeans came to North America, some 60 million Bison roamed the continent. By 1900 there were fewer than 1000. Today there are about 30,000 Bison in the United States.

The Bison once ranged over most of North America from the Atlantic almost to the Pacific, and from Mexico north into Canada. Today, free-roaming herds are found only in Yellowstone National Park in the United States and in Wood Buffalo National Park in Northwest Territories, Canada. Custer State Park in South Dakota has the largest fenced herd.

The Bison is the largest terrestrial animal in North America. The males are up to 6 feet (1.8 m) high at the shoulder, from 10 to 12.5 feet (3-3.8 m) long and weigh from 900 to 2000 pounds (400-900 kg). Females are about 5 feet (1.5 m) at the shoulders, 7 to 8 feet (21-24 m) long and weigh from about 800 to 1000 pounds (360-450 kg). The head is massive and broad, topped by short, curving, pointed horns in both sexes. The Bison has dark brown shaggy fur over the head, neck, shoulders and forelegs. The fur on the rest of the body is shorter and sometimes lighter brown in color. Its tail is long with a tuft of fur at the tip.

Bison move in herds, but the sexes stay in separate groups except at mating time. From time to time, smaller groups gather into larger herds. When they were at their peak population, Buffalo herds sometimes stretched as far as the eye could see across the plains. They will stampede if startled, a characteristic the Indians used to advantage by forcing them over cliffs. They are excellent swimmers and they can run at speeds between 30 to 35 miles per hour. When they were abundant, they would go on annual migrations of 200 miles or

Above: **Bison or American Buffalo.**
Top: **Bison in Yellowstone National Park.**

more. Those in Canada still make annual migrations of 150 miles or so between valleys and hill regions.

Neither nocturnal nor diurnal, Bison are active throughout the day. However, they are most active in the early morning and late afternoon. They also show increased activity on moonlit nights. They feed mostly on grasses and sedges and they will also eat lichens and horsetails. During the hot part of the midafternoon, they generally rest, chewing their cuds or dust-bathing. Dust-bathing or wallowing is done to repel insects and control itching. Males sometimes urinate in a dry dust wallow and cake themselves with the resulting mud as protection against insects.

Mating behavior begins in July when bulls rejoin the herd and start to fight for cows. The bulls stand about 20 feet from each other, lower their heads, and charge into each other at high speed. The heads, rather than the horns, collide. The charges are repeated until one of the bulls retires. They apparently suffer no serious injury during this fighting. Successful bulls form harems of from 10 to 70 cows. Mating takes place in August and September, and calves are born in May. The births are usually single, but twins occur occasionally. Covered with yellowish-red fur, the calf stands to nurse 30 minutes after birth and is ready to walk with its mother in a few hours. They are able to follow the herd when they are two days old.

The Bison was central to the life of the western Native Americans for meat, clothing and shelter. The destruction of the American Bison was started around 1830 as an official government policy aimed toward subjugating hostile Indian tribes through starvation. The railroads also contributed to the near extinction of the Bison. Some railroad building crews subsisted almost entirely on buffalo meat. After the railroads were built, hunting excursions were promoted and passengers shot Bison by the thousands from moving trains.

Today, the 30,000 remaining Bison are scattered in national and private preserves. Some of the herds, however, are thriving and, occasionally, they must be culled by controlled hunting to prevent the population from exceeding the food supply.

***Above:* Bobcats live only in North America.**

BOBCAT
(Felis rufus)

The most common wildcat in North America, the Bobcat's range extends in scattered populations over most of the United States and southern Canada. It is believed to be most abundant in the far West, and it is probably almost absent in most of the central and lower midwest and the San Joaquin Valley of California.

A tawny tan in the summer and somewhat gray in the winter, the Bobcat is some 2⅓ to 4 feet (.7-1.2 m) long and weighs from about 14 to 70 pounds (6.3-32 kg). The males are larger than the females. The legs have black horizontal bars, and the face has thin black lines that fan out into a broad cheek ruff. The short tail has a black tip above and is white on the underside.

A daytime hunter, the Bobcat hunts by hiding in a thicket, rock cleft or other concealed place where it waits for prey. A good climber, it also waits in trees, but less often than the Lynx. Usually well camouflaged by its spotted fur, it pounces on prey as they pass by. Tending to use the same hunting pathways repeatedly, it preys on snowshoe hares, rabbits, small rodents, porcupines and cave bats, but it very seldom eats carrion.

Although females can breed any time of the year, most come into heat in February or March. Litters, generally born in April or May, may contain one to seven kittens, but two or three is the usual number.

Bobcats may sometimes prey on domestic animals, particularly poultry if they can get into the poultry farm. Bobcats are hunted with hounds in some areas, and they are still killed for their fur, which is used for trim, although this practice is becoming more rare. The population appears to be stable or increasing in most areas.

The Bobcat is similar to the Lynx (*Felis lynx*) which has no spots and is a bit more gray. The Lynx generally ranges farther north.

CHAMOIS: see Goat, Mountain

CHICKAREE: see Squirrel, Red

CHIPMUNK, EASTERN; TAMIAS *(Tamias striatus)*

The range of the Eastern Chipmunk extends from southeastern Canada to all states of the continental United States east of the Mississippi except Florida, southern Mississippi, southwestern Alabama, southeastern Georgia, eastern North Carolina and most of South Carolina. West of the Mississippi it is found from northern Louisiana to Minnesota, and in the extreme eastern sections of Oklahoma, Kansas, Nebraska, South Dakota and North Dakota.

The Eastern Chipmunk is from 8½ to 19½ inches (21 to 30 cm) long and weighs from about 2¼ to 5 ounces (66 to 142 gm). It has reddish-brown fur on the upper parts and white fur below. White stripes, bordered by black, run down the sides. A wide dark band runs down the back and there are faint stripes on the face. Rounded ears top the pointed face.

The Eastern Chipmunk usually looks for food on the ground, but it can and will climb large trees such as oak and hickory to get acorns and nuts. It also feeds on seeds, fruits, insects, snails and other invertebrates, eggs, and meat of small vertebrates when it can get it. It often fills its cheek pouches with seeds and nuts. The Eastern Chipmunk stores food in storage burrows, constantly making trips between food-gathering sites and the burrow. It is often seen in the daytime scurrying about forest edges, brushy areas, stone walls and around houses. It chatters constantly with rapid trills and slow 'chuck-chuck' sounds. They are also noisy eaters, making snap-like sounds when they bite into nuts.

Chipmunks are true hibernators, holing up in underground burrows or tree limbs. They often wake up in midwinter and feed on their hoard of food. They also make use of the hoard when they awaken in the spring.

Eastern Chipmunks mate in early spring, and litters of three to five young are born in May. First-year females may have their first litters in July or August.

Above: **A Least Chipmunk.**

The Eastern Chipmunk may occasionally do some damage to flower gardens, but most of its activities are harmless. A fascinating, amusing animal to watch, it is sometimes kept as a pet. Many are taken by weasels, foxes, hawks and bobcats and some are killed by domestic cats and dogs.

Similar species are the Least Chipmunk (*Eutamias minimus*) and the Thirteen-lined Ground Squirrel (*Spermophilus tridecemlineatus*).

CHIPMUNK, LEAST *(Eutamias minimus)*

This small chipmunk is found from the southern Yukon and northeastern British Columbia, southeastward through southern Canada to northern Ontario and northern Minnesota, Wisconsin and Michigan's upper peninsula. In the eastern United States the range extends southward from central Montana to northern Arizona and westward to central Washington and Oregon and western California.

One of the smallest of the chipmunks, the Least Chipmunk is about 4 to 6 inches (10 to 15 cm) long and weighs from 1 to 2½ ounces (28 to 70 g). Color of fur varies with the region in which it is found, from a pale, yellowish-gray with tan dark stripes in dry areas to brownish-gray with black

the sides is orange-brown and the underside is grayish-white. The rather long tail is light brown above and yellowish below.

Adapted to live in a variety of habitats, the Least Chipmunk lives in sagebrush deserts, in coniferous forests and on rocky cliffs. The Least Chipmunk feeds on seeds, acorns, soft fruits such as berries, grasses, fungi, insects and other invertebrates. It may eat a small vertebrate such as a mouse. It readily climbs trees for acorns and nuts. Like most chipmunks, it stores food and hibernates in a burrow built under a stump log or rock. Its distinctive call is a series of high-pitched chirping notes.

Litters of two to six young are born in May.

The Least Chipmunk does no damage to crops or gardens, and is often seen in national park campgrounds.

Similar species include the Eastern Chipmunk (*Tamias striatus*), Townshend's Chipmunk (*Eutamias townsendii*), and the Redtail Chipmunk (*Eutamias ruficaudus*).

COTTONTAIL, EASTERN *(Sylvilagus floridanus)*

One of the most frequently seen rabbits in North America, the Eastern Cottontail is found in all of the United States east of the Mississippi except northern New England and northern New York. The range extends westward to the tier of states from Texas to North Dakota, eastern New Mexico, eastern Colorado and central Arizona, and southward into Mexico. In Canada it is limited to southern Ontario and small areas in southern Manitoba and Saskatchewan.

The Eastern Cottontail is from 1.2 to 1.5 feet (.35 to .45 m) long and weighs from 2 to 4 pounds (.9 to 1.8 kg). As is the case with most rabbits and hares, the does are larger than the bucks. The fur is grayish-brown and flecked with a little black on the upper regions and white below. The nape of the neck has a distinctive rust-colored patch.

Active from evening to late morning, Cottontails are often seen grazing in back yards in the early morning. They adapt to a range of habitats including open fields, woods, cultivated fields, gardens, thickets and brush piles. Strictly herbivorous, cottontails eat a variety of green vegetation in the summer and twigs and bark in the winter. Like other rabbits and hares, they eat quickly and return to a hiding place where they defecate soft green pellets of undigested food. These are later eaten and digested. During the

***Above:* A Nuttall's or Mountain Cottontail (*Sylvilagus nuttallii*) in Canada.**

day Eastern Cottontails hide in a burrow beneath a brush pile or in a slight cavity or depression in the ground. In the winter, they may make a network of runs under the snow when there is sufficient brush to support the snow. In very cold weather, they might hole up in an abandoned burrow of another animal. They sometimes play in groups on crusted snow.

The Cottontails usually move by hopping; however, they can move quickly in leaps of ten to fifteen feet (3 to 4.6 m). When pursued by a predator, they circle, often jumping sideways to break the scent trail. Sometimes they 'freeze' when threatened. When alarmed, they may thump the ground with their hindlegs, setting up vibrations that can be detected by other rabbits. If necessary, they will take to water, although they prefer not to get wet. Normally silent, they can give a piercing scream-like distress call. Does may grunt or purr while nursing, and the young squeak softly.

The breeding season runs from February to September. Males fight with each other and attempt to attract the attention of does with elaborate courtship rituals. A doe can produce three or four litters a year that can contain as few as one or as many as nine young. However, four to five young is the usual litter size. The doe makes a nest in tangled brush, high grass or other concealed areas. She digs out a depression in the ground and lines it with downy fur she bites off from her underside, mixed with dead grass and leaves. The biting off of fur also serves to expose the nipples, facilitating nursing. The doe hides the blind furless young under a layer of nest lining while she is away foraging, returning to them at dawn and dusk for nursing. The young rabbits are nursed for about two weeks, after which they may leave the nest. A doe is ready to mate again hours after giving birth. Without a high predation rate, cottontail rabbits would soon overpopulate their feeding areas, as indeed they do occasionally in isolated areas free of predators. The Eastern Cottontail is widely hunted and frequently consumed. The meat is considered to be excellent, but should be well cooked to avoid parasitic worm infestation, especially in warmer climates. These rabbits may also carry tularemia, a serious febrile disease. When dressing a rabbit it is sensible to wear gloves as a precaution against tularemia. Pelts, which are not particularly durable, are sometimes used for trim.

Similar species include the New England Cottontail (*Sylvilagus transitionalis*), the smaller Desert Cottontail (*Sylvilagus audubonii*), Swamp Rabbits (*Sylvilagus aquaticus*) and Marsh Rabbits (*Sylvilagus palustris*) and the larger Snowshoe Hare (*Lepus americanus*).

COYOTE
(Canis latrans)

***Above:* The Coyote or 'barking dog.'**

The range of the Coyote is known to extend over western Canada and all of the United States including eastern Alaska, and southward into practically all of Mexico. It is also found in southern Ontario and extreme southern Quebec. Although commonly associated with the West, Coyotes appear in Wisconsin, Michigan, Indiana, Ohio, Pennsylvania, Vermont and western Massachusetts.

Smaller than wolves, Coyotes are from 2 to 2.2 feet (.6 to .7 m) high at the shoulder, 3.5 to 4.3 feet (1.1 to 1.3 m) long, and weigh from 20 to 50 pounds (9 to 23 kg). The fur is a grizzled or reddish gray above, and somewhat buff below. The legs have a rusty-yellowish fur interrupted by a vertical line of black on the foreleg. The tail is bushy with a black tip.

Coyotes hunt at anytime of day, but are more likely to be active at dusk and night. They eat whatever they can catch, including rabbits, rodents, gophers, birds, snakes, lizards, frogs and insects. They have been known to take larger prey, such as deer and pronghorn, although the cooperative efforts of two or more coyotes are usually needed to bring down larger animals. If they cannot find animal prey they will eat fruit and carrion. They frequently hunt in pairs or groups of three or four, and one might chase the prey toward another coyote waiting in ambush. They also steal the prey of other carnivores through such tactics as waiting at one end of a burrow to pounce on an animal as it emerges after being flushed out by a predator such as a badger.

Well adapted to a variety of habitats, they can move for extended periods at speeds of 25-30 mph (40-48 kph) and as fast as 40 mph (65 kph) for short bursts. Experiments with tagged coyotes have shown they can travel distances of 400 miles (654 km) or more. They are strong swimmers and will readily go after prey in water.

The vocalizations of Coyotes are familiar night sounds in many parts of the United States, particularly the western states, and have passed into legend as the epitome of mournfulness and solitude in the great open spaces. This vocalization is a series of yelps and barks followed by a sustained howl that often terminates with a few short yaps. When one Coyote starts to vocalize, others usually join in, creating the cacophany of howls and barks that is a familiar night chorus in the West. The howling is believed to be a way of keeping members of a hunting band in communication.

Many Coyotes are believed to stay with the same partner for life. Coyotes mate from February to April. Litters consisting of one to twenty pups are born in April through May. The young are born in a nesting chamber which is typically part of a den located at the end of a five to thirty foot long tunnel. Dens may be abandoned badger or fox burrows enlarged by the female. If dens are not available, the female will have her pups in a variety of protected spots such as logs, culverts or caves.

The Coyote population continues to increase despite poisoning, trapping and the loss of their habitat to human settlement. They are trapped for their fur and killed by sheep ranchers and poultry growers who regard them as threats. Ecologists point out that while a Coyote will, indeed, take a domestic animal if it can, it is of greater benefit through killing rodents that prey on chicks and other young livestock. Suburban housing developments have encroached on the coyote's habitat on the fringes of cities such as Los Angeles and Phoenix.

DEER, MULE
(Odocoileus hemionus)

The most abundant deer in the west, the Mule Deer is found in the west coast states, all the Rocky Mountain states, North Dakota, South Dakota, Nebraska, Kansas, Oklahoma, Texas, Minnesota and Iowa, and in Canada from western Manitoba to the Pacific, and north to the Yukon and Northwest Territories. In Mexico it occurs in Baja California and in the north central states.

The Mule Deer is about 3 to 3.5 feet (.9-1.1 m) high at the shoulder and some 4 to 6.6 feet long (1.2-2.0 m). Bucks, which are larger than the does, weigh about 120 to 450 pounds (55-205 kg), while does weigh from about 80 to 150 pounds (36-68 kg). The upper and outer hair is reddish to yellowish-brown in summer and somewhat grayish in winter. A cream to tan color persists in all seasons on the undersides, throat patch, rump patch and inside of the ears. Bucks have equally-branching antlers with a spread of about 4 feet (1.2 m). The antlers are shed in February. The distinctive feature that gives the animal its name is the large, long ears. About 4.5 to 6 inches (11.5-15 cm) long, each ear can be moved independently. Mule Deer have a stocky body and sturdy-looking legs. Back and front legs move together, giving them a stiff, bounding gait.

In the summer, Mule Deer are active mostly in the mornings, evenings, and on moonlit nights. In the winter they are more likely to browse at midday in addition to morning and evening. They have a varied diet that includes both herbaceous and woody plants. Herbaceous plants are the more frequent summer fare, while woody plants comprise the winter browse. They also eat acorns and apples. Mule Deer that live in mountainous areas avoid heavy snow by migrating down the slopes in winter.

The Mule Deer usually moves in small groups, seldom gathering in large herds. A doe forms a small group consisting of fawns and yearlings. Doe groups tend to keep to themselves in areas spaced out from other groups. If does meet, they are likely to fight. These fights serve to maintain the feeding areas of each group. Bucks are generally solitary, but small groups may form just before and after the rutting season. Herding, when it does occur, tends to be limited to the winter.

During the mating season, which begins in November, the bucks establish small harems of does through fights and display threats. Fights are infrequent, consisting of crashing of antlers, with each buck trying to bring the other's head down. Serious injuries are rare, and the major hazard is the possibility of the antlers becoming locked, in which case both bucks will probably starve to death.

***Above:* A Mule Deer doe.**
***Right:* A Mule Deer buck.**

Fawns are born after a gestation period of six to seven months. Newly-bred does give birth to one fawn, while older does usually deliver twins. Fawns are concealed by the mother for the first month, after which the fawns follow the doe. Fawns may stay with the mother for two years, at which time they are sexually mature. The fawns seem to recognize the mother by sniffing at secretions of a gland that is located on each of the hind feet. Also present on the bucks, hairs around the glands stand erect when the bucks engage in aggressive behavior.

The Mule Deer, a popular game animal in the western states, is prized for its meat and as a trophy. In some areas, they damage crops and young timber. They are prey for coyotes, cougars and wolves. Bears and bobcats may take fawns or an occasional adult.

The Mule Deer and White-tailed

Deer (*Odocoileus virginianus*) are in the same genus and have many similarities. The White-tailed Deer does not have the long ears of the Mule Deer. Also, the Mule Deer's nose is considerably broader than the White-tailed Deer's, and it has a shorter, blunter face. The Black-tailed Deer is considered to be a subspecies of the Mule Deer by most biologists. It differs from the Mule Deer primarily in the brownish to blackish color of the tail.

DEER, WHITE-TAILED; DEER, VIRGINIA *(Odocoileus virginianus)*

The most familiar and abundant of the American deer, the White-tailed Deer ranges over almost all of the United States with the exception of most of California and sections of Arizona, Colorado, New Mexico and Nevada. It is also found across southern Canada and in extreme northern Mexico.

The White-tailed Deer is about 3 to 3.5 feet (.9-1 m) at the shoulder and some 4.5 to 7 feet (1.4-2.1 m) long. The bucks weigh from 200-300 pounds (90-136 kg) and the does weigh from about 150-250 pounds (68-115 kg). Tan to reddish brown in summer, and a darker grayish-brown in winter, the White-tailed Deer gets its name from its tail which is brown but is edged with white above and is all white below. When alarmed, the animal 'flags' its tail, exhibiting the white underside. The bucks have antlers with a forward main beam and several unbranched tines. Does do not usually have antlers. Antlers are shed shortly after the mating season.

Although it is primarily nocturnal, the White-tailed Deer can be active at any time of day, looking for food in woods, brushy areas, and land that has been cleared for farming. It eats a wide variety of plant foods, including nuts, aquatic plants and corn. In the winter its diet is mainly woody plants such as twigs and buds of trees. When food is scarce, it may strip trees of bark. Good runners and strong swimmers, they can move at speeds of 35 mph (56 kph) in short bursts. They usually take to cover when alarmed, rather than run great distances. Their vocalizations are limited to snorts when they are nervous and agitated.

Bucks and does move in separate herds through most of the year. In winter they gather in small herds of a male, a few does, and young. These herds may coalesce into groups of 25 or more individuals.

The White-tailed Deer breeds from late September to November. Fawns are born after a gestation period of seven months. First-time mothers usually give birth to one fawn, while subsequent births are usually twins, and triplets may occur if food is abundant. The doe hides the spotted newborn fawns in thickets for the first four or five weeks following birth, after which the young follow the

***Above:* A White-tailed Buck Deer.**

mother. The white-tail 'flag' helps the young to follow the doe in flight.

The White-tailed Deer was near extinction in some deforested areas of the northwest and midwest. However, the decline in predators such as mountain lions and bobcats, and restrictions on hunting have resulted in an increase in deer population in these areas. In some areas, particularly the east, the populations tend to exceed food supplies, and they become agricultural pests, raiding orchards and vegetable crops. Hunting is limited to specified seasons in most areas of the United States, although extended seasons and increased kill allotments are instituted from time to time in areas where there are overpopulation problems. Today the most abundant game animal in eastern North America, they are hunted for their meat (venison), which is considered to be excellent, and as trophies. Many are struck by automobiles and killed.

Two smaller subspecies are the Key Deer (*O.v. clavium*) of the Big Pine Key area of Florida and Coue's Deer (*O.v. covesi*) of the Arizona and New Mexico desert. These 'toy' deer seldom exceed 50 pounds (23 kg) in weight. An endangered species, the Key Deer was protected by the National Key Deer Refuge set up in 1961, and is making a comeback.

Similar species include the Mule Deer (*Odocoileus hemionus*), which tend to be heavier than White-tailed Deer, and which have rather long ears.

DOLPHIN, ATLANTIC BOTTLE-NOSED; PORPOISE; DOLPHIN *(Tursiops truncatus)*

The most commonly seen dolphin in the waters off North America, the range of the Atlantic Bottle-nosed Dolphin extends from Cape Cod southward into Central and South American waters.

The Atlantic Bottle-nosed Dolphin attains a length of about 12 feet (3.6 m) and weighs from 300 to 400 pounds (135 to 180 kg). The overall body color is gray, and is somewhat lighter on the underside than on the upper parts of the body. A pointed, swept-back dorsal fin is located about midway between the head and tail. The fish-like body ends in a horizontal tail or fluke.

The Atlantic Bottle-nosed Dolphin tends to swim in groups of 50 to 60 individuals. Capable of swimming at speeds as fast as 20 mph (32.5 kph), they are often seen swimming in the wake of ships, frequently leaping out of the water with the body in a graceful arch, in a maneuver called 'porpoising.' Feeding almost exclusively on fish, dolphins locate prey and find their way around in the underwater dark-

Above: **A female Dolphin with her young.**

ness by means of echolocation or 'sonar.' They emit high-pitched clicks and squeaks which bounce back to the dolphin as echoes, and are interpreted to give information on the size, speed, shape and location of everything in its environment.

Capable of remaining submerged for an hour or more, the dolphin must come to the surface to breathe. The animal breathes through the blowhole located on top of the head. Also used to produce vocalizations such as a high-pitched squeal, the blowhole is closed while the dolphin is submerged.

Breeding in spring and summer, dolphins seem to engage in almost constant sexual play. After a gestation period of about 12 months, a single young is born underwater. Immediately pushed to the surface, the young can submerge for less than a minute when first born, but slowly increases the amount of time it can remain underwater. Nursing underwater, the nursing periods are, of necessity, short and frequent at first, but increase in duration and become less frequent as the young dolphin grows. Young dolphins remain with the mother and nurse for as long as two years, although they are able to eat fish by the time they are about three months old.

The trained dolphin of the seaquarium shows that the Atlantic Bottle-nosed Dolphin has a high degree of intelligence and they are thought to be able to communicate with each other in a highly developed language. It does well in captivity, learns quickly and seems to enjoy the company and attention of humans.

Classified as a toothed whale, there are more than 30 species of dolphins in the world, of which about ten are commonly seen in North American waters. These include the Pacific Bottle-nosed Dolphin, the Atlantic and Pacific White-sided Dolphin, the Common Dolphin, the Spotted Dolphin, and the Striped Dolphin. Large numbers of dolphins in the Pacific drown in nets of tuna fishermen when they become trapped going after the tuna. New net designs and techniques have reduced the numbers of dolphins lost in nets, enabling fishermen to usually keep within kill limits set by regulation.

DOLPHIN, STRIPED: see Dolphin, Atlantic Bottle-nosed

ELK, WAPITI *(Cervus e. canadensis)*

Before European settlers came to North America, the range of the Elk extended over most of what is today the United States and southern Canada. Now, the Elk is primarily a western species. It is found in Colorado, Utah, Wyoming, Montana, Idaho, western Washington and Oregon, northwestern California, extreme northern Arizona and New Mexico, Vancouver Island, southwestern Alberta, eastern British Columbia, central Saskatchewan and southern Manitoba. Isolated groups are found in central Arkansas, south central California, Nevada, Oklahoma, South Dakota and Michigan. Isolated remnants of the once large eastern population are seen in a few eastern states, particularly Pennsylvania and Virginia.

One of the largest of the deer, the bull Elk is from 4½ to 5 feet (1.4-1.5 m) high at the shoulder, some 6.5 to 10 feet (2-3 m) long, and weighs from 600 to 800 pounds (270-362 kg). The cow Elk weighs about 300 pounds less (136-226 kg) and is proportionately smaller. The fur is brown, lighter on

***Above:* A herd of Elk or Wapiti.**

top and darker on the underside. The tail and rump hair are a lighter yellowish-brown. Males have dark brown hair on the throat. The most notable feature on the bulls is the large antlers. When mature, these antlers, which are shed every spring, have six tines or branches and have a beam length of 5 feet (1.5 m) or more.

Elk are active mostly in the night hours, particularly at dusk and dawn. Although they are primarily grazers, they will feed on practically any kind of plant. They are strong animals capable of running speeds of 35 to 40 miles per hour (56-64 kph) and swimming in strong currents, but they can move through the forest almost silently in a graceful, pacing trot. In the winter they graze at lower altitudes, often in thick forest. In the summer, they move to higher, open pastures. They mark their territories by stripping bark off seedling trees. Cows strip the bark with their teeth, and bulls use the base of their antlers, chin and muzzle.

Elk make a variety of vocalizations. The adults constantly make snorting and grunting noises, and the calves squeal. Cows call their calves with a sound similar to a horse's neigh. During the rutting season bulls, which are the most vocal of American deer, make their 'bugle' or 'whistle' call. It begins as a low bellow, changing almost immediately to a loud, high-pitched noise that can be described as a whistle or scream. The 'bugle' is both a challenge to other bulls and a signal of domination over cows. Cows also whistle, but their vocalizations are not as loud as those of the bull.

Cows and young move in separate herds until the mating season, when they are joined by bulls. Running from August to November, the rutting or mating season reaches its peak in October and November. During the mating season, the males vocalize frequently, particularly with the bugle call. They also roll in mud puddles to coat themselves with mud and urinate on piles of verdure which they then pitch onto their backs with their antlers. Bulls engage in mating fights which consist of banging their antlers together. Although serious injuries are rare, some do occur, and a few are fatal when antlers lock. Bull Elks assemble harems that usually number six to eight cows. However, harems as large as 60 cows have been observed.

The cows leave the herd after a gestation period of 255-275 days to find a place to give birth. An Elk usually gives birth to one calf weighing from 25 to 40 pounds (11-18 kg). However, twins are not uncommon. The cow and calf rejoin the herd about a week after the birth, and the calf suckles for about nine months. As the young bulls grow older, they spend less and less time with the cow-dominated herd, and eventually leave to join herds of bulls. Cows stay with the female herds and are ready to breed at about age 2½. Elk have a lifespan of about 14 years.

The Elk was once more of a plains animal than it is today. They were, however, killed in large numbers by cattle ranchers who considered them to be pests that reduced cattle grazing areas. They were also hunted for their meat and antlers. Many were killed to satisfy a market for 'elks' teeth,' the upper incisor teeth, which were used as watch-fob charms.

The Elk is similar to other American deer, but is much larger than other North American deer species, with the exception of the Moose (*Alces americana*). The Moose has a large overhanging snout, and large, flattened antlers which give it an appearance quite distinctive from that of the Elk.

FOX, ARCTIC
(Alopex lagopus)

The Artic Fox is found in northern and western Alaska, northern Canada including the Arctic Ocean Islands, southward through the northern and eastern parts of the Northwest Territories, northeast Alberta, northern Manitoba, northern Quebec around Hudson Bay and northern Labrador.

The Arctic Fox is about a foot (.3 m) high at the shoulder, 2½ to 3 feet (.8 to .9 m) long, and weighs from 10 to 12 pounds (4 to 5 kg). Variable fur color is one of the many adaptations of the Arctic Fox to life in the Arctic regions. In the summer the fur is gray or bluish-brown, and white on the undersides. In the winter the coat thickens and the new hair that grows in is white. There

Above: **An Arctic Fox in winter colors.**

is also a rare blue phase, which is a pale blue-gray in the winter and a darker blue-gray in the summer. Short legs and ears cut down on heat loss, and fur on the foot pads provides traction on ice and insulates against cold.

The Arctic Fox's varied diet includes rabbits and hares, rodents, birds, birds' eggs, carrion, fish and berries. Occasionally they may be able to take a young seal or sea lion. They sometimes follow polar bears, eating the remains of their kills. In the summer, the Arctic Fox hunts and eats enormous amounts of food even when it is not hungry. In the winter, surplus food is stored by freezing it in holes dug in the ground or by hiding it under rocks. When food becomes scarce in the northern parts of its range, it may migrate southward.

Litters that can contain as many as 25 kits, but usually number six, are born in April and June. Cared for by both parents, the young go out on their own around August.

Many Arctic Foxes, particularly the blue phase, are trapped for their pelts. The Arctic Fox is raised commercially for fur on islands off Alaska.

The Arctic Fox is similar to other foxes such as the Gray Fox (*Urocyon cinereoargenteus*) and Red Fox (*Vulpes vulpes*), but has shorter ears and legs as well as distinctive color phases depending on the season.

FOX, GRAY
(Urocyon cinereoargenteus)

The Gray Fox is found in the eastern United States westward through most of Texas, Oklahoma, the Dakotas, Kansas, Nebraska, New Mexico, Arizona, Colorado, southern Utah and Nevada, western Oregon and most of California.

The Gray Fox is some 1⅛ to 1¼ feet (.3 to .4 m) high at the shoulder, from 2⅝ to 3⅓ feet (.8 to 1.1 m) long, and weighs from 7 to 13 pounds (1.8 to 6 kg). It has grizzled gray on most of the upper parts of its body, reddish fur on the underside and the top of the head. The tail is black fur along the top and the throat has a patch of white.

Above: **A juvenile Gray Fox.**

The Gray Fox does most of its hunting at night; however, it is often seen in the daytime. Its varied diet includes rabbits, rodents and other small mammals and insects. An excellent climbing ability enables it to include birds in its diet. Plant foods include apples, persimmons, grapes, cherries, pokeweed berries and grasses.

In the winter it may hole up in a den. Dens are usually small caves, hollow logs, trees and rock piles. However, if necessary, it will dig a den, either from scratch or by enlarging an abandoned woodchuck burrow.

Mating takes place in February, and litters that usually consist of two to four kits, but may contain as many as seven, are born in March or April. The male sometimes helps to feed the young, but he does not go into the den with them. The young Gray Foxes are ready to hunt on their own when they are four months old.

Related species include the Insular Gray Fox (*Urocyon littoralis*), the Red Fox (*Vulpes vulpes*), and the Kit Fox (*Vulpes macrotis*).

FOX, INSULAR GRAY: see Fox, Gray

FOX, KIT: see Fox, Gray

FOX, RED
(Vulpes vulpes)

The Red Fox is found in Alaska, practically all of Canada except for an area in Alberta and Saskatchewan and in most of the United States, except Florida, the coastal regions of North and South Carolina and Georgia, the western portions of Texas, Kansas, Nebraska and South Dakota, the eastern parts of Montana, Wyoming and Colorado, Arizona, coastal and southern California, eastern and western New Mexico (the range extends into central New Mexico). The Red Fox was originally a European species brought here to provide a hardier breed than the native gray fox for fox hunting. Present populations are a mixture of the Red Fox and the native gray fox which has expanded its range over the past 200 years.

The Red Fox is from 1¼ to 1⅓ feet (.4 to .5 m) high at the shoulder, from 3 to 3¼ feet long (.9 to 3.7 m) long and weighs from 7.5 to 15 pounds (3.4 to 7 kg). The hair is a rusty-red color above and white on the underside. The long bushy tail has a white tip, and the backs of the ears, lower legs and feet are black.

A shy, nervous animal, the Red Fox hunts mostly at night. It eats whatever it can find or catch, including both animal and plant foods. Prey includes small mammals, such as rabbits and rodents, insects and crayfish. It consumes a wide variety of plant foods including apples, cherries, corn, various berries, acorns and grasses. It caches surplus food in snow in the winter and in leaves or dirt in the summer.

The Red Fox mates from January to March. Shortly after mating, the pair sets up a den, usually in an area of sparse ground cover on a slight rise that provides a view of the surrounding area. The Red Fox den is commonly an enlarged abandoned burrow of a woodchuck or badger. Spare dens are built to which the kits are moved if the vixen perceives a danger. Litters are born from March to May, and may contain as many as 10 kits, but 4 to 8 is the usual number. For the first few days after birth, the male brings food to the maternity den; both share in this duty later. Kits emerge from the den at about age one month, and play at the den entrance. At first, the kits eat partially digested meat regurgitated by the parents, advancing later to live prey brought by the mother. At age four months, the kits can take care of themselves and leave the den. The vixen and the male split up around August, and remain solitary until the next early winter when they find new mates.

Red Foxes do not usually retire to dens in the winter, nor do they migrate. On a cold, windy day, a fox might curl up in the snow, protecting its nose and foot pads with its tail.

Red Foxes are both bred and trapped for their pelts, which are among the most widely used in the fur industry. In many areas, particularly where poultry growing was a big industry, there was a bounty on foxes. Today, most have been abolished.

***Above:* An adult Red Fox.**
***Below:* A pair of playful Red Fox Kits.**

Red Foxes take few chickens today, mainly because it is impossible to break into a modern poultry farm.

Similar species are the Coyote (*Canis latrans*), Gray Fox (*Urocyon cinereoargenteus*), and the Arctic Fox (*Alopex lagopus*), all of which lack the white tip at the end of the tail.

GLUTTON see Wolverine

GOAT, MOUNTAIN
(Oreamnos americanus)

The Mountain Goat is found in mountainous regions of northwestern Montana, Idaho and Washington, much of British Columbia, southwestern Alberta and southeastern Alaska. The mountain goat lives in mountainous regions above timber line.

Mountain Goats are from 3 to 3.5 feet (.9 to 1.1 m) high at the shoulder, are from 4 to about 6 feet (1.2 to 1.8 m) long, and weigh from about 100 to 300 pounds (45 to 135 kg). The males tend to be somewhat larger than the females. The Mountain Goat is a compact, chunky, short-legged animal with yellowish-white fur that is long and shaggy in the winter and shorter in the summer. Both sexes have a 'beard' that is about five inches (13 cm) long, and sharply tipped backward-curving black horns. The beard grows from the neck rather than the chin, as it does in true goats. The horns are about a foot (.3 m) long in the male and some 9 inches (23 cm) in the females.

Throughout the year the Mountain Goat moves up and down the mountains with the season, migrating to lower altitudes in the winter. The hooves have a hard outer rim surrounding a rubbery inner surface, providing excellent traction.

Most of the mountain goats' feeding activity is in the morning and evening or on bright moonlit nights. On a warm day, they may bed down during the late morning and afternoon. Their feeding follows a seasonal pattern. In the summer, they eat grasses, sedges, and other warm-weather plants and in the winter, moss, lichens and year-round woody plants.

***Above:* A Mountain Goat foraging.**

Mountain Goats form small herds of separate sexes until the mating season which extends from November to January. Males do not usually engage in fights over females. The skull bones and horns are fragile compared to those of other hoofed mountain animals such as bighorn sheep. Butting battles would likely result in injury. They do adopt threatening postures. Males sometimes mark or 'claim' a female by rubbing it with a musky oil that is secreted from a gland at the base of the horns. Kids are born in May and June. The usual number is one, but twins and triplets can occur. The kids stay with the mother until she gives birth again.

Mountain Goats trot with apparent ease across ledges, rocky outcrops, cliffs and ledges. Although they are remarkably sure-footed and able to move in areas that many predators could not negotiate, they do sometimes fall to their death. Many are killed by avalanches and rock slides. The major predators are mountain lions at lower altitudes and golden eagles in the peaks. Predators usually go after kids, in deference to the sharp horns and hooves of the adults.

While referred to as a goat, the Mountain Goat is not classified as a true goat by biologists. Mountain goats belong to the group called the goat antelopes. They are more closely related to the chamois, a deer-like animal of Europe, than to goats.

GOPHER, NORTHERN POCKET
(Thomomys talpoides)

A western species, the Northern Pocket Gopher is found in southwestern Canada, southward into North Dakota, South Dakota, the Rocky Mountain states as far south as northern Arizona and New Mexico, eastern Washington and Oregon, and northern Nevada.

The Northern Pocket Gopher is from 6 to 9 inches (15 to 23 cm) and weighs from $2\frac{3}{4}$ to $4\frac{3}{4}$ ounces (78 to 135 g). Varying in color from brown to yellowish-brown to gray, the fur color often approximates the color of the soil in which it lives. Built for a life of burrowing in the soil, this and other pocket gophers have thick bodies, relatively short, almost naked, tails, short necks and short fur that can lie forward or backward to facilitate movement in burrows. Large, yellow incisor teeth are always exposed in the front of the mouth, even when the mouth is closed. The lips remain closed when the gopher is gnawing or digging to keep dirt and debris out of the mouth. Pocket gophers get their name from large fur-lined external cheek pouches that extend from the cheek to the shoulder.

Like all pocket gophers, the Northern Pocket Gopher digs burrows. They dig burrows close to the surface for gathering food, and deeper ones for storing food. The major cutting tools are the long incisors. These large teeth grow throughout the life of the animal; if not, they would soon be worn down to nubs. The forefeet are used like shovels, and the hindfeet push the dirt behind the body.

The Northern Pocket Gopher moves through the surface burrows, foraging for roots, tubers and stems which it cuts and pulls down into the burrow. Able to obtain all its water needs from the food it eats, it does not have to go to the surface to drink. The Northern Pocket Gopher eats some food immediately and stores some in its pouch. From time to time, it goes down to the deeper tunnels and stores plant material in food chambers. The Northern Pocket Gopher empties the pouches by squeezing them forward with its forepaws, turning them inside out and cleaning them and then pulling them back into place.

It is a solitary animal, and a fight usually occurs if two Northern Pocket Gophers meet. While they do not hibernate, they retreat to deeper burrows and sleep during very cold weather. In the spring, males start to seek females, and male-female meetings during mating season are almost the only time when encounters of these gophers do not result in fights. One to two litters a year of one to two young are born. Young gophers start to dig burrows while they are still being cared for by the mother. When they are about two months old they disperse.

Pocket Gophers damage crops, particularly alfalfa and root crops. Their burrow mounds interfere with harvesting, but they also aerate the soil and bring subsoil to the surface.

The Northern Pocket Gopher is similar to most other pocket gophers including the Plains Pocket Gopher (*Geomys bursarius*) and the Western Pocket Gopher (*Thomomys mazama*).

GOPHER, PLAINS POCKET
(Geomys bursarius)

The Plains Pocket Gopher is found in the prairie areas of the midsection of the United States from northeastern Texas northward through Oklahoma, Kansas, Nebraska, southern and eastern South Dakota, eastern North Dakota, western Louisiana, western Arkansas, northeastern Missouri, Iowa, most of Minnesota, western Wisconsin, central Illinois, southeastern Wyoming, northeastern Colorado, south central New Mexico and a small area in northern Indiana.

The Plains Pocket Gopher is from 7¼ to 14 inches (18 to 35 cm) long and weighs from 4 to 12½ ounces (113 to 355 g). Those that live in the northern part of the range are larger than those in the southern parts. The fur color varies from light brown to black depending on the color of the soil. Most of the black gophers are in Illinois.

Like all pocket gophers, the Plains Pocket Gopher digs burrows. The burrows of this pocket gopher are in sandy loam, loam soils, grasslands and sometimes along railroad tracks. In the summer, the burrows are only about an inch from the surface; in the winter, they are dug considerably deeper. Food is gathered in the upper burrows and stored in chambers, usually at lower levels. Food is stored temporarily and transported in the gopher's long cheek pouches or 'pockets' which are emptied by the gopher pushing its forepaws against them.

Mating begins in early spring. After a gestation period of 18 to 19 days, litters of one to eight young are born. The nests are in deeper underground tunnels in the north and may be in large surface mounds in the south. Those that live in the south may have two litters or more a year.

Plains Pocket Gophers can be quite destructive to crops and lawns.

Similar species include the Northern Pocket Gopher (*Thomomys talpoides*) and the Valley Pocket Gopher.

GROUNDHOG:
see Woodchuck

HARE, SNOWSHOE
(Lepus americanus)

The Snowshoe Hare is found in almost all of Alaska except the extreme north and the extreme southeastern areas, southward into almost all of Canada except extreme southern Ontario, the extreme north and the Arctic Ocean Islands southward to New England except Connecticut and eastern Massachusetts, New York, Pennsylvania and the Appalachian area in the east, centrally to northern areas of Michigan, Wisconsin, Minnesota and North Dakota, and in the west southward from Canada to Washington, most of Oregon, northern Idaho, western Montana, western Wyoming, the central areas of Utah and Colorado, northeastern California and north central Arizona.

One of the smaller hares, the Snowshoe Hare is from 1¼ to 1¾ feet (.4 to .5 m) long and weighs from 2 to 4 pounds (.9 to 1.8 kg). The name of the Snowshoe Hare stems from the large hindfeet that are from 3¾ to 6 inches long (9.5 to 15 cm) and have fur-covered soles. The fairly long ears are from 2½ to 3¼ inches (4.6 to 8 cm) long. In the summer months, the fur is a dark brown and the tail is grayish-brown with a white underside. In most parts of the range the fur turns white in winter. Those in Washington and Oregon stay brown all winter, and a small group in the Adirondack Mountains of New York are black. The changing of fur color is caused by the periodic increase and decrease of the hours of daylight as the seasons change. Although the white fur provides protective coloration in winter snow, there are times when the changing and molting of fur does not synchronize with snowfall and melt. The first snow may come before the Snowshoe Hare's change to white is complete, and the animal will be exposed to predators. The Snowshoe Hare seems to know that it is vulnerable at these times and spends more time hidden in protective cover.

Active mostly at night, the Snowshoe Hare spends the day resting in a

***Above:* A Snowshoe or Varying Hare, in its winter coat.**

protected place such as a hollow log, abandoned woodchuck burrow or a form. It feeds on grasses, berries and green vegetation in the summer and browses on bark and conifer buds in the winter. It is one of the few rabbits and hares that will eat meat if it is available, and it often steals bait from traps. It often takes a dust bath, frequently in the dusting wallows of grouse. If forced to flee from a predator, it will run in a wide circle of several acres in leaps of twelve feet (3.6 m) or more, often leaping sideways to confound its pursuer. If it can, it will take protective cover rather than run. A good swimmer, if necessary it will go into water to escape a predator, although it does not like to get wet.

Two to three litters of one to six young are born a year. Like most hares, the young are capable of running about within hours of birth. The population of Snowshoe Hares appears to peak at nine to ten year intervals and then drop and slowly increase to another peak. The population fluctuations may be related to those of predators, particularly the lynx, whose population peaks and valleys follow the Snowshoe Hare's by a year. Another factor that may contribute to the fluctuation is overcrowding that produces stress which interferes with reproductive activity.

The Snowshoe Hare is a major food of a number of predators, including, the lynx, fox, weasel, mink, wolverine, hawks and owls. An important game animal, it has been known to damage gardens and newly planted trees.

Similar species include the Arctic (*Lepus arcticus*) and Tundra Hares.

HARE, TUNDRA: see Hare, Snowshoe

HARE, VARYING: see Hare, Snowshoe

HARE, WHISTLING: see Pika

JACK RABBIT, BLACK-TAILED *(Lepus californicus)*

The most abundant and widespread of the jack rabbits, the Black-tailed Jack Rabbit is found in much of the southwestern United States from Texas, northward to Nebraska and westward to Oregon. It is also found in parts of Arkansas and Missouri, southern South Dakota, south-eastern Wyoming, and central Washington. It is also found in much of Mexico.

The Black-tailed Jack Rabbit is from 1½ to 2 feet (.45 to .6 m) long and weighs from 4 to 8 pounds (1.8 to 3.6 kg). The fur on the upper parts of the body is sandy to buff gray and speckled with black. The fur on the underside is white, and the tail has a black stripe on the top that extends forward to the rump. This hare has very large ears that range from 5 to 6 inches (13 to 15 cm) or more and large hindfeet that are from 4¼ to 5¾ inches (11 to 14.5 cm).

A hare rather than a rabbit, since it bears its young above ground rather than in the underground burrow used by the rabbit, the Black-tailed Jack Rabbit is the most commonly encountered of the large jack rabbits in the barren prairies and meadows of the southwestern United States. Active mainly at night, its summer fare includes a variety of plant foods, particularly alfalfa which seems to be a favorite. In the winter, it eats dried woody vegetation. More gregarious than most hares and rabbits, it feeds in loosely associated groups.

The large ears of the Black-tailed Jack Rabbit help it to pick up sounds from some distance away. It can often be seen standing still, moving its ears to catch sounds. It usually moves by hopping, rather than walking. Hops of five to ten feet (1.5 to 3 m) are a normal pace; if fleeing from a predator, the Black-tailed Jack Rabbit is capable of twenty foot (6 m) leaps and speeds of 45 mph (72 kph) for short bursts. At moderate hopping speeds, every fourth or fifth leap is unusually high, apparently to give it a better view of the surroundings. It 'flashes' the underside of its tail when pursued, apparently to alert other jack rabbits of the danger.

If necessary, it will take to water to avoid a predator, swimming in a dog-paddle with all four feet. Black-tailed Jack Rabbits are preyed on by foxes, coyotes, snakes, owls and hawks.

Mating occurs at any time of year, but is more frequent in warmer weather. One to four litters of one to eight young are born a year. As is the case with all hares, the young are born fully furred and with eyes open. The mother deposits her young in scattered forms to reduce the chances of all being lost to a predator. She stays away by day, but returns at night to nurse. The young can take care of themselves at age one month.

The Black-tailed Jack Rabbit is considered to be a nuisance by cattle ranchers because it eats vegetation that can be used as cattle graze. The meat is considered to be fair.

Above: **The Black-tailed Jack Rabbit.**

Below: **A White-tailed Jack Rabbit in the Grasslands National Park, Canada.**

JACK RABBIT, WHITE-TAILED
(Lepus townsendii)

The White-tailed Jack Rabbit is found from southern Alberta, Saskatchewan and Manitoba, southward into the United States from Wisconsin westward to central Washington, southward to the Sierra Nevada mountain region and westward to Kansas and Iowa.

The White-tailed Jack Rabbit is from 1⅞ to 2⅜ feet (.6 to .7 m) long and weighs from 5½ to 10 pounds (2.5 to 4.5 kg). The fur is buffy gray above, and the underside, tail, and the black ears have white edges. The large ears are 3¾ to 5 inches (9.5 to 13 cm) long. In most parts of the range, except for the southern areas, the fur turns white or pale gray in the winter.

Active mainly at night, the White-tailed Jack Rabbit feeds on clover, grasses and other green vegetation in the summer and browses on twigs, buds and other dried plant material in the winter.

Rather than walking or running, the White-tailed Jack Rabbit normally hops in twelve to twenty foot leaps (3 to 6 m). It can move at a sustained speed of about 35 mph (56 kph) and can attain short bursts of 45 mph (72 kph). If necessary, it will go into water and swim in a strong dog paddle.

Mating takes place in late April and May. Bucks fight viciously for does, kicking furiously with their hind feet and biting. Litters of one to six young are born after a gestation period of 30 days in loose vegetation rather than prepared nests. Growing rapidly, the young White-tails are independent at age three or four weeks.

The White-tailed Jack Rabbit is considered to be a fine game animal with meat that is more tender than that of other hares

LEMMING, SOUTHERN BOG
(Synaptomys cooperi)

The only lemming with an extensive range in the continental United States, the Southern Bog Lemming is found in the Maritime Provinces, southern Quebec and Ontario, southward into the United States from New England to Virginia, eastern Kentucky, the Blue Ridge areas of Tennessee and North Carolina, westward to Kansas and northward to Minnesota.

The Southern Bog Lemming is from 4½ to 6 inches (11 to 15 cm) long and weighs from ¾ to 1¾ ounces (21 to 50 g). The fur is brown above and silvery gray below.

It spends much of its time in runways and burrows about 6 inches (15 cm) below the ground. Spherical grass nests are built in underground chambers and on the surface. More likely to be found in meadows and forests than in bogs, the major components of this lemming's diet are clover and grass, supplemented with algae and fungi.

Above: **A Mountain Lion or Cougar, the most widely found North American cat.**

In the southern part of the range, the Southern Bog Lemming breeds throughout the year, while in the northern range, breeding is more likely to be limited to the summer months. Three or more litters of two to six young a year may be produced. The Southern Bog Lemming has wide fluctuations in population. Similar species include Voles and the Northern Bog Lemming (*Synaptomys borealis*).

LION, MOUNTAIN; CATAMOUNT; COUGAR; PUMA
(Felis concolor)

Although the number of Mountain Lions in North America today is only a fraction of what it was a hundred years ago, it is still the most widely distributed cat on the continent. The Mountain Lion is mainly a western cat with a range that extends from British Columbia and southern Alberta south through the Rocky Mountain states and west coast areas and western Texas. Isolated populations are also found in the eastern United States in Louisiana, Alabama, Tennessee, Arkansas and Florida.

One of the larger cats in the western hemisphere, the Mountain Lion is from 6 to 7 feet (1.8-2.1 m) long and weighs from 70 to 280 pounds (32-128 kg). Its fur is a tawny yellow above and white with a buff overlay below. The long tail is tipped with black, and the rather small head is tipped with rounded, small ears. Vocalizations include hisses, grunts and a hair-raising mating call that sounds like a woman screaming in pain. They apparently vocalize only when excited, alarmed or agitated.

Except at mating time, Mountain Lions are solitary. They stalk prey in a variety of habitats, including mountains, forested hills, swamps, semi-arid, subtropical and tropical areas. While most cats are mainly nocturnal, the Mountain Lion, although it prefers to hunt at night, is active at any time of day. They are excellent climbers and can jump more than twenty feet and can run as fast as a deer for short distances. It may travel miles in a single night in search of game. Its preferred prey are larger mammals such as deer. It will, however, take whatever is available including beaver, porcupines and other rodents, rabbits, raccoons, birds and insects. Mountain Lions usually take prey by advancing on it slowly to a distance from which they can pounce or run it down. They will also lie in wait for other animals in tree branches and usually leap on the victim's back, biting into the neck.

Females come into heat at any time of the year. Pairs form when a female is in heat, and during the breeding period of about two weeks the pair will hunt and sleep together, mating frequently. After a gestation period of about 90 days, the female seeks out a protected place such as a cave, thicket or crevice where she gives birth to one to six kittens. Young lions may stay with the adults for as long as two years before they become independent.

Designated as an endangered species, many were killed by ranchers and farmers who considered them a threat to livestock. In some areas there was a bounty on them. Although a Mountain Lion will take a domestic animal, such as a sheep or cow, if it can, such kills seldom result in serious economic loss. On rare occasions, however, a single Mountain Lion might, for unknown reasons, kill an entire herd of sheep in one night. Older lions generally prey on old and diseased deer. Mountain Lions are rarely dangerous to man. Latin-American legend holds that their nature is playful and friendly and that a child may sleep safely out of doors when they are nearby.

The jaguar (*Felis onca*) is similar to the Mountain Lion, but is extremely rare in the United States. The Jaguarundi (*Felis yagouaroundi*) is a similar, but much smaller, species.

LYNX
(Felis lynx)

The Lynx is found in deep forested areas in Alaska and most of Canada. Its range also includes areas of Washington, Oregon, Idaho, Montana, Wyoming and Colorado. Small populations exist in extreme northern New England, northern New York, Michigan and Wisconsin.

A stocky cat with short legs, the Lynx is about 2½ to 3½ feet (.8-1.1 m) long and weighs from 10 to 40 pounds (4.5-18 kg). Males are somewhat larger than females. It has a buff or tawny coat sprinkled with black hairs and the underside is a darker cinnamon-brown color. The cheeks have a paler cast than other parts of the body and the ears have long, black tufts. Whitish hairs hanging from the throat form a forked beard. The large feet are covered with fur.

The Lynx hunts mostly at night. A frequently-used hunting tactic is to wait in a tree and pounce on prey passing below. The wide, furred feet enable the Lynx to move swiftly and silently through snow. Its favorite food appears to be the snowshoe hare, one of the few animals that can move through the snow with greater speed than the Lynx. The population of the Lynx and snowshoe hare fluctuate and peak every nine to ten years in a cycle that is a classic ecological study. The Lynx also eats small rodents, birds, an occasional weakened deer or other hoofed mammal, and carrion. It rarely takes a domestic animal. The Lynx is prey for wolves, mountain lions, and hunters, who kill them for their fur.

Mating takes place in March and April. Kittens, usually in litters of two, are born between May and July. The kittens remain with the mother until the following spring. A similar species is the Bobcat (*Felis rufus*), which is more tannish and has a longer tail.

MARMOT, YELLOW-BELLIED; MOUNTAIN MARMOT; ROCKCHUCK; YELLOW-FOOTED MARMOT
(Marmota flaviventris)

Found mostly in higher elevations up to 11,000 feet (3,350 m) of the northwestern United States, the range of the Yellow-bellied Marmot includes central British Columbia, southward to Idaho, most of Washington and Oregon, most of Nevada, eastern California, western Montana, most of

***Below:* The Lynx is a strong swimmer.**

Above: **A Yellow-bellied Marmot.**

Wyoming, central Utah and north central New Mexico. The range also includes a small area of southwestern South Dakota.

The Yellow-bellied Marmot is from 1½ to 2⅓ feet (.45-.7 m) long and weighs from 5 to 10 pounds (2.2 to 4.5 kg). The fur on the upper parts of the body is yellowish-brown, while that on the underside is yellow. The fur on the neck is light, and seems to separate the body fur from the darker fur on the face.

The Yellow-bellied Marmot feeds on a variety of vegetation such as grasses, herbs and alfalfa found in the rocky slopes, valleys and foothills that comprise its habitat. It constructs a den under a rockpile, crevice or other sheltered place. Not usually venturing very far from its den, the Yellow-bellied Marmot dashes into it when alarmed, and emits several chirps and whistles once it is safely inside.

A true hibernator, the Yellow-bellied Marmot puts on a layer of fat in preparation for the winter period of inactivity. It remains in hibernation from around August to February or March. In the mountains of New Mexico, it may start hibernating in August. Some may also estivate in the summer.

Litters of three to six young are born from March to April. They emerge from the den when they are about a month old.

The Yellow-bellied Marmot can do serious damage to crops of alfalfa. Abundant in Yellowstone and other National Parks, they are interesting animals to observe.

MINK
(Mustela vison)

The Mink's range extends through all of Alaska, most of Canada except the extreme northern area and the Arctic Ocean Islands, southward to include most of the continental United States except the southwestern part of the country from western Texas westward to southern New Mexico, Arizona, most of Utah and Nevada, and southern California.

The Mink is 1.1 to 1.4 feet (.3 to .4 m) long and weighs 1½ to 3 pounds (.7 to 1.3 kg), and the females are 1 to 1.2 feet (.3 to .36 m) long and weigh from 1¼ to 2⅓ pounds (.6 to 1.1 kg). Minks that live in the wild have fur that ranges from brown to black with spots of white on the chin and throat.

The Mink spends much of its time near rivers, lakes, marshes, ponds and other bodies of water, preying mostly on muskrats but also including rabbits, mice, chipmunks, fish, snakes, birds, turtles and an occasional chicken stolen from a poultry farm in its diet. They also den near water, frequently in the abandoned burrow of a muskrat. Minks run down their prey, pouncing on them and biting them in the neck. Kills are usually eaten on the spot, but surplus is often cached in the den. Like many weasel-like animals, they give off a foul-smelling musk from their anal glands. It marks its hunting territory with the musk, and generally fights any other mink that violates it.

Minks mate in January and February, and litters of three to six young are born in the spring. Males fight viciously during the mating season, mate with several females, but eventually stay with one female for the duration of the breeding season.

The lustrous fur of the Mink is among the most valuable in the fur

Below: **The Mink is prized for its fur.**

industry. Very few Mink, however, are trapped today compared to the past. Most fur comes from mink raised on ranches where mink are bred to produce a fairly wide range of colors.

Similar species are the Marten (*Martes americana*) and the Black-footed Ferret (*Mustela nigripes*).

MOLE, EASTERN; COMMON MOLE
(Scalopus aquaticus)

The Eastern Mole is found in southern New England, southern New York, eastern Pennsylvania and all other states east of the Mississippi River with the exception of most of West Virginia and part of Wisconsin. West of the Mississippi, it is found from Louisiana and most of Texas to Nebraska, Iowa and southern Minnesota.

The Eastern Mole is from 3½ to 8½ inches (9-22 cm) long and weighs from 3 to 5 ounces (85-142 g). It has broad paddle-shaped forefeet that have turned-out palms. The tail is very short and the toes are slightly webbed.

Like all moles, the Eastern Mole is a burrower. Its progress through the soil is aided by a body that is almost without a neck indentation. The velvety fur has no grain, an adaptation that allows it to move backwards with as much facility as it moves forward. A narrow pelvis enables it to change direction quickly in the narrow confines of its tunnel, sometimes by somersaulting.

Most active at dawn or dusk, the Eastern Mole spends practically all the time underground. It burrows or 'swims' through the soil, shoving aside the earth with alternate sweeps of its forefeet and packing the dirt behind it with its hindfeet. Its main food is earthworms, but it also feeds on insect grubs, insects and vegetable matter.

Moles are solitary except at mating time when males go into tunnels other than their own to seek a female. The males might even come to the surface in their search for a mate. Mating takes place in March and litters of four to five young are born some six weeks later in a maternal nest chamber lined with dried plant matter. The young are sexually mature at age ten months.

Moles make deep tunnels and subsurface tunnels. It is the latter which are often marked by ridges and by six to eight foot high (1.8-2.7 m) molehills that mar the appearance of lawns and gardens. Sometimes the burrow damage can be repaired by simply walking over the ridges. Thousands of moles are killed every year with poison and traps, but moles may more than make up for the damage they cause by eating Japanese beetle grubs.

The Eastern Mole is not particularly liked by many homeowners who are concerned about the appearance of their lawns. The burrowing activities of these animals do, indeed, damage lawns and golf courses but they also help to control harmful insects.

Similar species include the Hairy-tailed (*Parascalops breweri*) and Star-nosed Mole (*Candylura cristata*).

MOOSE
(Alces americana)

The largest deer in the world, the Moose is found in most of Canada and Alaska. In the rest of the United States, it is limited to Maine, extreme northern Minnesota, the northern fringe of Michigan's Upper Peninsula, and the continental divide area of the Rocky Mountains in Idaho, Montana, Colorado, Utah and Wyoming.

The Moose is about 6½ to 7 feet (2-2.3 m) high at the shoulder, and from 7 feet to 9 feet (2.1-2.9 m) long. Males weigh from 900 to 1400 pounds (408-635 kg) and females weigh from 700 to 1100 pounds (318-500 kg). The Moose has dark brown hair and grayish brownish fur over its long legs. The shoulders have a high humped appearance and the face has a pendulous, curved muzzle and a large, rather long, dewlap hangs under the chin. The most distinctive characteristic of the male is the large, flattened antlers. The antlers, which have small prongs at the edges, have a usual spread of 4 to 5 feet (1.2 to 1.5 m), but spreads of 6¾ feet (2.05 m) have been observed. The antlers are shed between December and February.

Moose are most active at night, but can be seen at any time of the day. Generally solitary in the summer, they come together in small groups during the winter. However, groups of several individuals gathered near streams and lakes to feed on water plants and willow trees are common

Below: **A common or Eastern Mole burrows for earthworms, its favorite food.**

summer sights. In addition to water plants, Moose eat twigs, buds, and the bark of balsam, aspen, birch, maple and other trees. The Moose spends a great deal of time in or near water. Some of their favored foods are water plants such as water lilies, and they commonly submerge to escape from swarms of black flies. Wallowing in mud puddles to acquire a protective coating of mud is another method of protection from those flies. Moose are drawn to salt licks and will eat salty earth near their mud wallows.

They can move quietly at speeds up to 35 mph (56 kph). Strong swimmers, they can swim up to two hours at a stretch. During the rutting season, males are quite vocal, frequently emitting a loud roaring bellow that can be heard for some distance. The cow's call is softer, ending in a cough-like sound. Hunters often try to lure the Moose within shooting range by imitating the mating calls of both cow and bull through cupped hands or a horn.

The seasonal migration of Moose consists of moving up and down mountain slopes. The winter herds pack down the snow, which makes it easier for them to move and find food. During the mating season, which runs from about mid-September to late October, a bull will tend to stay with one cow for about a week before going on to another. Mating battles between the bulls are rare; a threatening posture will usually cause one Moose to retreat.

Calves are born in May after a gestation period of about 245 days. Only one calf is usually born, but twins are fairly frequent. Calves are able to walk and swim along with their mothers when they are two weeks old.

The Moose is highly prized as a game animal, both for its meat and as a trophy. It was almost exterminated by hunters in the eastern United States before it was protected by law. Although they usually avoid contact with humans, they can be dangerous. Cows with a calf are extremely protective. Bulls in rut are unpredictable and will charge almost anything for no apparent reason.

Similar species are elk (*Cervus canadensis*) and other deer. Elk and caribou most closely resemble the Moose in size and structure, but both lack an overhanging snout and flat antlers.

***Above:* Moose are usually seen near water.**
***Below:* A juvenile Moose.**

MOUSE, COTTON
(Peromyscus gossypinus)

The Cotton Mouse is found in the south central United States from eastern Texas to all of Louisiana, southern Arkansas, western Tennessee, southwestern Kentucky, all of Mississippi, Alabama and Florida, southern Georgia, and northward along the Atlantic coast to Maryland.

The Cotton Mouse is from 6¼ to 7¾ inches (16 to 20 cm) long and weighs ¾ to 1¾ ounces (21 to 50 g).

At home in a variety of habitats, including wooded areas, brushy areas,

Above: **A Deer Mouse, one of a variable species.**

rocky areas, swamps and beaches, the Cotton Mouse is an excellent tree climber and swimmer. Omnivorous and noctural, it feeds on seeds, nuts, fruits, insects and other invertebrates.

The female has one or more litters of one to seven young a year.

Similar species include the smaller White-footed Mouse (*Peromyscus leucopus*) and Deer Mouse (*Peromyscus maniculatus*), the Oldfield Mouse (*Peromyscus polionotus*) and Golden Mouse (*Ochrotomys nuttalli*).

MOUSE, DEER
(Peromyscus maniculatus)

One of the most widely distributed of the mice, the Deer Mouse is found from central Alaska southeastward to the Yukon, the southwestern Northwest Territories and all of the rest of Canada except northern Quebec and the Arctic Ocean Islands. In the United States it is found in all states west of the Mississippi except Louisiana, eastern and southern Texas and southern Arkansas. It is found in all states east of the Mississippi River except the deep South, southern Tennessee, eastern North Carolina and Virginia, southeastern Pennsylvania, Delaware, eastern Maryland, Connecticut, Rhode Island and eastern Massachusetts.

The Deer Mouse is from 4¾ to 8¾ inches (12-22 cm) long and weighs from ⅜ to 1⅛ ounces (11-32 g). It is grayish to reddish brown above and white below.

Occurring in many varieties the deer mouse exists in two main forms – the Prairie and the Woodland. The prairie form is the smaller of the two and is found mostly in the Midwest. The woodland form is found mostly in the North, although there is considerable overlapping of the populations.

This Mouse has a varied diet that includes seeds and nuts, berries and other small fruits, insects, centipedes and ground fungi. The Deer Mouse may store seeds and small nuts in logs and other protected areas. The prairie form commonly nests in cultivated areas, eating wheat and corn seeds and caterpillars.

The Woodland Mouse digs small burrows in the ground and is likely to build nests in hollow logs.

The mating season varies with the latitude, but generally occurs from February to November. Two to four litters of one to eight young are born a year. Females are ready to breed at age five to six weeks.

Deer Mice, even those living in cultivated fields, seldom eat enough to cause a discernible loss and may be beneficial by eating caterpillars. Similar species are the White-footed Mouse (*Peromyscus leucopus*) and Cotton Mouse (*Peromyscus gossypinus*).

MOUSE, FIELD:
see Vole, Meadow

MOUSE, HOUSE
(Mus musculus)

Originally an Asian species, the House Mouse came to North America with Spanish explorers and *conquistadores* who landed in Florida in the early 1500s. It was also brought to northern North America in English and French ships. Today, the House Mouse is found throughout the continental United States, southern Canada and in Mexico.

The House Mouse is a grayish-brown above and, unlike most other mice, it does not have white fur on its underside. While the fur on the underside is somewhat lighter than that on the upper side, a difference in color is not particularly noticeable it is about 2½ to 4 inches (6-10 cm) long and weighs ⅝ to ¾ ounces (18 to 21 g).

The House Mouse has become well adapted to areas of human habitation, including houses and other buildings, and in cultivated fields Almost completely omnivorous, the House Mouse will eat practically anything, but it seems to be more selective in its food when it lives in cultivated fields and other outdoor areas. It is more likely to eat weed seeds, caterpillars and insects than crop plants in fields, where it can actually be of benefit to farmers. However, when it gets into buildings such as houses, barns and storage houses it becomes very destructive, eating whatever it can get and contaminating what it doesn't eat with urine and droppings. The droppings may contain eggs of parasitic worms that can infest humans. These mice chew on almost anything, damaging furniture, walls, floors and sometimes starting fires when they gnaw on electrical wires. Millions of dollars worth of stored grains are destroyed or contaminated by the House Mouse.

In the wild they build individual nests, but live in groups in a way that is neat and ordered compared to their ways inside houses. The nests have separate areas for eating and waste elimination. They groom each other constantly, particularly on the head and back.

Much of the reproductive and other behavior of the House Mouse is governed by fluctuations in population. Breeding throughout the year, the House Mouse has an enormous reproductive potential. The gestation period is only about three weeks and a female can produce several litters a year. If the population becomes too dense in a particular area, many females become sterile. If they are in areas of abundant food, such as a cultivated field, the rate of reproduction increases. When the crop is harvested, the mice disperse, often going into buildings.

In 1926-1927, there was a massive population explosion of the House Mouse in the central valley of California, believed to have been caused by excessive sport killing of snakes and birds of prey, the major predators of the House Mouse. The normally retiring mice swarmed into towns, biting people and crowding into houses and other buildings.

The House Mouse has been selectively bred to produce albino strains, the 'white mice' that are used in research laboratories and sold as pets.

Similar species include the White-footed Mouse (*Peromyscus leucopus*) and the Harvest Mouse.

MOUSE, PLAINS POCKET: *(Perognathus flavescens)*

Limited in range to the western United States, the Plains Pocket Mouse is found from southern Minnesota, southeastern North Dakota and most of Iowa, southwestward to Nebraska, most of Kansas, eastern Colorado, western Oklahoma, the Texas panhandle and extreme southeastern New Mexico.

The Plains Pocket Mouse is from 4½ to 5 inches (11.4 to 13 cm) long and weighs from ¼ to ⅜ ounces (7 to 11 g). The soft fur is yellowish-buff sprinkled with black. External cheek patches are adapted for carrying food.

Inhabiting sandy, sparsely vegetated plains, the nocturnal Plains Pocket Mouse forages on seeds and leafy plants, eating some on the spot and storing some. Obtaining its water needs from the seeds it eats, it does not need to drink water. These small rodents dig systems of tunnels with separate chambers for food storage sleeping and nesting. During the day it plugs up the main entrance to the nest but leaves smaller entrances open.

Several litters a year (more in the southern range) of one to eight young are born in burrow nests. Similar species are the Silky Pocket Mouse (*Perognathus flavus*) and Merriam's Pocket Mouse (*Peromyscus merriami*).

Above: **Musk oxen forage for woody plants in the winter.**

Below: **A Plains Pocket Mouse.**

MUSK OX *(Ovibos moschatus)*

Musk Oxen have roamed the North American arctic region in great numbers, but they are now comparatively rare and their range is limited to Alaska, Greenland and some islands of the Arctic Ocean. Overhunting was a major factor in their decline. Although their appearance is that of a true ox, they are more closely related to goats and sheep.

The bull Musk Ox stands 4 to 5 feet (1.2-1.5 m) tall and is usually 7 to 8 feet (2.1-2.4 m) long. Bulls weigh between 700 and 900 pounds (317-707 kg), but as with most of the ruminants, cows are smaller. The Musk Ox has short, stocky legs and a proportionately heavy body which looks even heavier because of the animal's humped shoulders and thick hair. Its shaggy dark-brown coat is curly and matted over the shoulders and straight on the rest of the body, not unlike that of the bison. The bull bears hollow horns joined at the base of the forehead and curving down, outward and up to a sharp point. The Musk Ox takes its name from the pungent odor it emits.

Musk Oxen feed on grass, willows, lichens and other small plants of the tundra. They will fight courageously and powerfully if attacked by wolves or bears, their chief natural enemies. Bulls are especially aggressive during the mating season in September.

The female Musk Ox is ready to breed at the age of two years, and calves are born in late May or early June, sometimes relatively few of them to a herd. Observed populations at Spitsbergen in the Arctic Ocean appear to reproduce themselves very slowly.

Efforts to keep and breed Musk Oxen in American zoos have had very limited success due to their nonresistance to indigenous sheep and cattle diseases.

MUSKRAT
(Ondatra zibethicus)

The extensive range of the Muskrat includes all of Canada and northern Alaska except the extreme northern regions, and all of Florida, most of California, southern Arizona and New Mexico, central Texas, southern Alabama, and Georgia, extreme southeastern North Carolina, and a small area in west central Utah.

A fairly large rodent, the Muskrat is from 1⅓ to 2 feet (.4 to .6 m) or more in length and it weighs from 1 to 4 pounds (.45 to 1.8 kg). The roundish, somewhat vole-shaped body is covered with dense, glossy fur that is a rich, dark brown above and somewhat lighter on the sides, becoming paler on the undersides and almost white on the throat. The tail is from 7 to 12 inches (18 to 30 cm) long, scaled, and is flattened vertically, tapering to a point. The partially webbed hind feet are larger than the forefeet.

This aquatic rodent spends most of its time in and around bodies of fresh water such as rivers, lakes, ponds, canals and fresh, brackish or saltwater marshes. Excellent swimmers, they propel themselves forward and backward with their partially webbed hindfeet, using the vertically flat tail as a rudder. Able to remain submerged for several minutes at a time, the mouths close behind the incisor teeth, allowing them to chew underwater without swallowing water.

Muskrats build houses of aquatic plants, particularly cattails, in marshy areas and along riverbanks. In marshes they may be built on supports such as piles of roots. The houses are as much as 8 feet (2.4 m) in diameter and some 5 feet (1.5 m) high. The houses constructed in marshes usually contain only one nesting chamber, while those made along river banks may have several chambers and many underwater entrances. Those that live along rivers also construct burrows in the river banks. Feeding platforms, made of cut vegetation, are also built in water or on ice. The Muskrat keeps the houses and feeding platforms scrupulously clean, and constantly repairs them and makes additions. Houses usually contain only one individual or family, although two or more nonrelated individuals may occupy the same house more or less amicably except at breeding time.

The Muskrat's varied diet includes aquatic vegetation, some terrestrial plants, crayfish, frogs, fish and freshwater clams in areas where they can be found. Food is usually dragged to the feeding platforms for consumption.

The mating season begins in September. Female Muskrats can produce one to five litters of one to eleven young a year. At age two weeks, the young muskrats have fur and can swim and dive. They are driven away by the mother when they are a month old.

MYOTIS, INDIANA; SOCIAL MYOTIS
(Myotis sodalis)

One of the more commonly seen bats in the Midwest, the range of the Indiana Myotis extends from New England eastward through New York, and Pennsylvania to Missouri, northward to the shores of Lakes Michigan and Erie and southward to northern Virginia, Tennessee, Arkansas and the northern parts of Alabama, Mississippi and Georgia.

The Indiana Myotis is from 2¾ to 3⅝ inches (7 to 9 cm) long and weighs about ¼ ounce (7 g). It has a distinctive pinkish brown color.

The name of this bat alludes to the way they sleep or hibernate in neat, tightly packed rows. The bats hang from the roof of a cave or other den in tightly packed clusters, usually one row deep. The noses and lips of the bats appear to be a continuous line of pink in rows of bats formed with almost military precision.

Mating takes place shortly after the hibernating groups form in October. They emerge from hibernation in the spring, and the young are born in June. Some groups break up during the summer months.

Similar species include the Little Brown Myotis (*Myotis lucifugus*), the Mississippi Myotis and the Gray Myotis (*Myotis grisescens*).

MYOTIS, LITTLE BROWN; LITTLE BROWN BAT
(Myotis lucifugus)

The Little Brown Myotis, one of the most commonly seen bats in the United States, is found in southeastern Alaska, most of Canada except the northern regions, and southward through almost all of the United States except most of Texas, southeastern New Mexico, the southern fringes of Georgia, Alabama, Mississippi and the southern parts of Nevada and California.

The Little Brown Myotis is from 3⅛ to 3¾ inches long (8 to 9 cm) and weighs no more than ⅛ to 1½ ounces (3.5 to 14 g). It is a glossy brown above and somewhat buff in color on the undersides.

In the summer, the Little Brown Bat lives in nursery colonies in secluded parts of buildings. In the winter those that live in the East hibernate in caves and mines. Others migrate hundreds of miles.

The Little Brown Bat leaves its roost at dusk to hunt insects which it captures on the wing, with the aid of a well-developed echolocation sense.

Above: The Little Brown Myotis or Little Brown Bat is the most common American bat.

Swarms of bats can sometimes be seen at dusk around cave or mine entrances.

The nursery colonies begin to form in April or May. The young, which are born anytime from May to July, are nursed constantly for the first two to three days after birth, except when the mother is out hunting. A Little Brown Bat might carry its offspring in flight if it is disturbed. The young bat holds on to a teat with its mouth, and tucks its hindlegs under the opposite armpit of its mother. The young bats are independent when they are a month old.

The guano, or droppings, of the Little Brown Bat was once a valuable fertilizer. Although many people have a revulsion toward this often seen bat, the Little Brown Bat is completely harmless and it helps to control insect pests.

Similar species include the Indiana or Social Myotis (*Myotis sodalis*), the Cave Myotis (*Myotis velifer*) and the Mississippi Myotis.

OPOSSUM, VIRGINIA
(Didelphis marsupialis)

One of the better-known behaviors of the opossum has become part of our language. 'Playing possum,' which has come to mean pretending inaction in order to deceive an adversary, alludes to the opossum's rolling over and appearing to be dead when confronted by a predator. This is probably not a voluntary act, but a catatonic involuntary reaction to stress. Often the predator is confused enough to leave the opossum alone. 'Playing possum' is but one of the many adaptations that have helped this rather primitive mammal to survive in today's world. The Opossum, the only marsupial in North America, is not only thriving, but seems to be increasing in some areas of its range.

The Opossum is found throughout the eastern United States except Maine, west to Colorado and Texas, north to Minnesota, and south into Mexico. It is also found in Pacific coastal areas and in some parts of Arizona and New Mexico.

The Opossum is from 25 to 40 inches (80-100 cm) long, and weighs from about 4 to 15 pounds (1.9-6.8 kg). In the north, the Opossum tends to be a grizzled white along its back with white hairs covering black-tipped fur along its belly. In the south they are brownish to blackish. The head, throat and cheeks are whitish, and the black ears have white to pinkish tips. It has short legs with an opposable thumb-like first toe on the hind feet. The Opossum's prehensile tail is long and naked. A good tree climber, it uses both its feet and tail for climbing.

Mostly nocturnal, the Opossum feeds on insects, frogs, birds, eggs, snakes, earthworms and small mammals. It also eats plant food, particularly apples, corn and persimmons when they are in season. It also eats carrion. Many are killed by automobiles when they are attracted to highways by the remains of other animal highway casualties.

Active throughout most of the year, the Opossum does not hibernate.

Below: A juvenile Opossum.

However, it does hole up in a den in extreme winter conditions. The den can be a hollow tree, log, or an abandoned burrow of another animal. Although it generally 'plays possum' when cornered, it will fight or at least threaten in some instances. The Opossum has 50 teeth, more than any other North American land mammal.

A female Opossum might have as many as three litters a year. When she is ready to deliver, she builds a nest of leaves in the den. After a gestation period of only 12-13 days, the one to fifteen (average about ten) tiny offspring, which have been described as 'living embryos,' are born. A typical litter of newborn Opossums can fit comfortably into a teaspoon. Only about ½ inch (1.3 cm) long, and weighing only some 1/15 of an ounce (2 g), they climb through the mother's hair and enter the brood pouch through a vertical slit. Eleven to 13 milk nipples are inside the pouch. Each of the young takes a nipple in its mouth – which it is then unable to release for some weeks – and continues its development. Sometimes there are not enough nipples for all the young oppossums, and those who do not obtain a nipple do not survive. The size of the young increases some 1000 percent in a week. They continue to take milk for about nine weeks, and at age four weeks they stick their heads out of the pouch; by age five weeks they leave the pouch for short periods of time. When they are eight weeks old they can take care of themselves. They are ready to breed at age one year and they live for about seven years.

ORCA: see Whale, Killer

OTTER, RIVER
(Lutra canadensis)

The River Otter is found in Alaska, all of Canada except the extreme northern regions, and is unevenly distributed through all of the continental United States except southern Texas, the extreme southern areas of Arizona and New Mexico, eastern Utah, southern California and most of Nevada.

***Above:* A River Otter (*Lutra canadensis*).**

The River Otter's elongated body is from 3 to 4 feet (.9-1.2 m) long and it ranges in weight from 10 to 30 pounds (4.5-13.6 kg). The males are larger than the females. It has dark brown fur that appears to be a shiny black when wet. The underside is a lighter brown, the throat region of many individuals is silvery-gray, and prominent, and there are white whiskers on the face. It has webbed feet and a long tail that is very thick at the base and gradually tapers to a point.

***Below:* Sea Otters are playful animals.**

As its name suggests, the River Otter spends most of its time in and around bodies of fresh water such as rivers, lakes and ponds. It does, however, wander away from the water from time to time. Although its main food is fish, crayfish and other aquatic animals, it also eats small terrestrial mammals such as mice. When catching fish alone, it snaps them up quickly. Often, pairs of Otters work together to drive fish into a blind inlet. Fish are eaten both on land and in the water while the Otter floats on its back. It hunts primarily by day in areas away from human habitation, and it tends to be more nocturnal in populated areas.

Its many adaptations for life in the water include a lithe, streamlined body, webbed feet, a tail that acts like a rudder and ears and nostrils that can be closed off to keep out water. A strong, graceful swimmer, the Otter can move at high speeds both submerged and on the surface, and can remain submerged for minutes at a time. It can swim backwards almost as rapidly as it can swim forward and it also moves well on land.

OTTER, SEA
(Enhydra lutra)

The Sea Otter is found in Pacific coastal waters, no more than a mile from shore, in rocky shallows and kelp beds from California to Alaska.

The Sea Otter is from 4 to 5 feet (1.2-1.5 m) long and weighs from 30 to 80 pounds (13.5-36 kg). Males are larger than females. The body fur is dark brown and that of the head and back of the neck is yellowish to grayish. Whiskers grow on the face. All feet are webbed, and the hind feet are constructed like flippers.

The Sea Otter stays in the water most of its life, going ashore only when there are severe storms. An excellent, strong swimmer, the Sea Otter uses its feet, body and tail for propulsion. The Sea Otter dives for much of its food, and can stay submerged for 4 to 5 minutes.

The Sea Otter feeds on fish, abalone, sea urchins, crabs, clams, mussels and other sea life, which it eats while floating on its back. Often, when it comes to the surface with a shelled sea animal such as a clam or abalone, it also brings a rock, places the rock on its chest and smashes the shell against it.

Sea Otters appear to be fun-loving creatures, and will often interrupt a meal to play and frolic with other Sea Otters. It also plays with other animals such as seals and sea lions. From time to time, it will stand in the water, shade its eyes with its paws and look out over the sea for predators. It sleeps in the water, wrapping a piece of kelp around its body as a tether. The Sea Otter has no insulating blubber to protect it from cold, depending instead on air trapped in its fur. If the fur is slicked down by oil from a spill, the animal could die from exposure.

Sea Otters mate in the water, and the young are born at sea in May or in June. The female usually has one kit and twins occur rarely. The mother floats on her back while the young nurses, naps and plays on her belly. If alarmed, the mother will tuck the kit under her arm and submerge.

The Sea Otter was hunted extensively for its fur almost to the point of extinction until the trade was stopped by an international treaty in 1911. It made a slow comeback, but was thought to be extinct in California until 1938, when a herd was discovered near Carmel. The population in that area is thought to be about 1000 today, while much larger populations exist off the southern coast of Alaska and in the Aleutian Islands.

PIKA; CONY; WHISTLING HARE; PIPING HARE
(Ochotona princeps)

The Pika is found in scattered populations from southern British Columbia southward to central Washington and Oregon, northeastern California, and southward from the Montana-Idaho border to central Utah and Colorado, central Nevada and northern Arizona. It lives in steep boulder-covered hillsides, rock slides and other rocky areas in mountains at elevations from 8000 to 13,500 feet (2440 to 4115 m).

The Pika is from 6¼ to 8½ inches (16 to 22 cm) long and weighs from 3½ to 4¼ ounces (9 to 11 g). The fur is brownish over the entire body. Shaped somewhat like a small guinea pig, the Pika has small rounded ears and no visible tail.

Unlike most of the related rabbits and hares, the Pika is a gregarious, very social animal that lives in large, constantly vocalizing colonies. They communicate with chattering sounds and shrill nasal bleats. When it makes a sound, it jerks its body forward.

Active by day, the Pika consumes a wide variety of green plants, eating some as it is found, while some is stored. In the late summer, the Pika scurries back and forth from feeding sites to boulders near its den where it spreads vegetation in 'haystacks' to dry. Some stacks may contain a bushel or more of vegetation. The stacks consist mostly of grasses and sedges, but they also include thistle, sweetgrass, stonecrop and fireweed. Stacks are constantly tended to enhance drying and to protect them from rain. When the vegetation is well dried, it is stored in the Pika's den, which is usually under a rock. Each Pika seems to maintain its own food stack territory.

***Above:* A Pika, Cony or Whistling Hare.**

The Pika remains active throughout the harsh winter of its high altitude habitat. Well protected by its long thick fur, it feeds on lichens and its hoard of dried vegetation.

Two litters a year of two to five young are born in May or June and July or August.

Often seen in national parks in the mountain states, Pikas do not live in built-up areas, and do not eat crop plants or compete with cattle for graze. The Collared Pika (*Ochotona collaris*) is similar and considered the same species by some biologists.

PORCUPINE
(Erethizon dorsatus)

The range of the Porcupine includes most of Alaska and Canada and extends southward from Canada into New England, New York, most of Pennsylvania, and small areas of Virginia and West Virginia, from central Canada southward into northern Michigan, Wisconsin and Minnesota, and from western Canada into Washington, Oregon, most of California, the Rocky Mountain states, almost all of Arizona and New Mexico, western portions of North Dakota, South Dakota, Nebraska, and Kansas, southwestern Texas and the north Texas panhandle.

One of the larger rodents, the Porcupine is 2 to 3 feet (.6 to .9 m) long and weighs from 7 to 40 pounds (3.2 to

18 kg). Its name is derived from the French for 'spiny pig.' The large, chunky body has a high-arching back and the legs are short. The body is covered with long hairs in the front, and sharp quills are on the rump and tail. Hair color varies from black or brown in the eastern part of the range to a yellowish color in the West. There are small fleshy knobs on the soles of the feet, which have long curved claws.

Active mainly at night in wooded areas, the strictly vegetarian Porcupine forages alone for a variety of plant foods including leaves, twigs, and leafy green plants such as skunk cabbage and clover. Its major winter food is the tender inner bark (cambium) of trees, which it gets to by chewing off the outer bark. It has a particular liking for salt, which it often satisfies by eating salt spread by road crews and by eating sweat-soaked wooden tool handles. The Porcupine does not hibernate, although in particularly cold weather it may hole up in a burrow or rocky bluff, sometimes with other Porcupines. During the day, it may rest in a number of sheltered places such as burrows, hollow logs or hollow trees, but it is more likely to be at the top of a tree, which it reaches in a slow but steady climb.

Not a fast mover, the Porcupine moves in a slow waddling walk. For protection against enemies it uses its quills, if it cannot take the preferred action of retreat or climbing a tree. A Porcupine has 30,000 quills on its body. Modified hairs, the quills are solid at the base and at the sharp tip and hollow through much of the shaft. The quills are loosely attached to a layer of muscle that lies just beneath the skin. Contrary to a rather venerable myth, a Porcupine cannot throw its quills. However, its method of delivering quills to an enemy is more than effective. The Porcupine erects the quills, lowers its head, and strikes the enemy with its tail, embedding the quills in the enemy's body. When the Porcupine retracts its tail, the loosely attached quills come free from the Porcupine and remain in the enemy. The victim's body heat causes small

***Above*: The Porcupine.**

barbs in the quills to expand, embedding them firmly in its flesh.

How much damage the quills do to the vicim is determined by their location. If they are in the mouth the victim may not be able to eat. They may enter the eyes and blind the victim. In almost all instances, the quill wounds become infected and fester. A few predators, notably the fisher and other weasel-like animals, bobcat, and coyote are adept at flipping the Porcupine over so that the unquilled underside can be attacked. However, even these animals often receive serious, sometimes fatal, quill injuries.

Mating takes place in October and November, at which time the Porcupines become very vocal, emitting groans, squeaks and grunts. One young Porcupine is born in May or June after a gestation period of seven months, which is rather long for a rodent. The young have quills when they are born. However, since the quills are soft and the young are in placental sacs and are born head first, the mother Porcupine is not injured during the birth.

PORPOISE: see Dolphin, Atlantic Bottle-nosed

PRAIRIE DOG, BLACK-TAILED *(Cynomys ludovicianus)*

The range of the Black-tailed Prairie Dog extends southward from extreme southern Alberta and Saskatchewan through eastern Montana to north-eastern Texas and includes the eastern portions of Wyoming, Colorado and New Mexico, and the western parts of North Dakota, South Dakota, Nebraska, Kansas and Oklahoma. The range also includes a small area of southeastern Arizona and extreme northern Sonora and Chihuahua in Mexico.

One of the larger ground squirrels, the Black-tailed Prairie Dog is from 1⅛ to 1¼ feet (.3 to .4 m) long and weighs from about 2 to 3 pounds (.9 to 1.4 kg). The fur is a pink-brown above and buff-white below. The thin, almost hairless tail has a black tip, a unique feature, which gives the animal its name.

One of the most gregarious of mammals, the Black-tailed Prairie Dog constructs extensive elaborate networks of tunnels, burrows and underground chambers in its short-grass prairie habitat. Seldom, if ever, going very far from the burrow entrances, the Black-tailed Prairie Dogs live in 'towns' extending over 100 or more acres, populated by several thousand individuals. Two hundred years ago, one town in the Texas Panhandle measured 250 miles long and 100 miles wide. Conical mounds of earth around the exits and entrance holes serve to keep out flood water and to provide 'observation posts' at which the animals sit on their haunches and scan the area for danger. 'Towns' are divided into smaller territories called 'wards' which are made up of family groups called 'coteries,' consisting of a male, one to four females and young up to age two years. Ward members recognize each other by touching noses and giving each other a side-wise 'kiss.' This is actually a contact of the incisor teeth for identification purposes. There is also a great deal of mutual grooming.

From the entrance the Prairie Dog digs the burrow extending down to a depth of about four or five feet (1.2-1.5 m) and then constructs lateral tunnels that serve as turn-arounds and listening posts. The entrance shaft is extended to depths of as much as 14 to 15 feet (4.25 to 4.5 m) from which a long horizontal 'hallway' tunnel is

Above: **A female White-tailed Prairie Dog.**
Left: **A Black-tailed Prairie Dog.**

built. Several nests and excrement chambers are built along the 'hallway.' The Prairie Dog covers its excrement with dirt and when the chamber is filled, it is closed off and the prairie dog digs a new one. Nesting chambers are lined with dry grass and the entrance and exit mounds, built at different heights, set up ventilating air currents in the burrows. Prairie Dogs cooperate in the building of burrows, and there is considerable interlinking of burrows. Prairie dogs go through recognition rituals when they meet in burrows, sometimes engaging in brief scuffles.

Black-tailed Prairie Dogs are very vocal. It is believed that the animal has nine separate calls, each of which has its own meaning. Calls include the shrill dog-like bark which gives the animal its name, squirrel-like chirps, chatters and snarls. The alarm call is a chirp, followed by a wheeze, to the accompaniment of a tail flick. Other Prairie Dogs pick it up and then dive into their burrows. The jump-yip is an interesting display consisting of the prairie dog giving a wheezy whistling 'yip' as it jumps up with its head thrown back and forelegs raised. It is thought to be an 'all-clear' signal.

The mating season runs from February to March. After a gestation period of about a month, one litter of four to five young a year is delivered in a nesting chamber. When they are about six weeks old, the young emerge from the burrow with the mother. They leave to take care of themselves at about age ten weeks, and are fully grown at age ten months.

The Prairie Dog population increased sharply after the bison was almost eliminated and larger tracts of western prairie were used for farming and raising cattle. Since Prairie Dogs were formidable competitors of cattle for prairie grasses, cattle raisers embarked on extensive extermination campaigns, primarily with poisons. In arid regions, valuable irrigation water can be lost to drain-off into burrows. However, studies have indicated that balanced populations of Prairie Dogs can actually improve rangeland and many ranchers make efforts to preserve some Prairie Dog towns on their acreage.

PRAIRIE DOG, WHITE-TAILED *(Cynomys leucurus)*

Restricted to a small range in the western United States, the White-tailed Prairie Dog is found at high elevations in western Wyoming, northwestern Colorado and east central Utah.

A stockily-built squirrel, the White-tailed Prairie Dog is from $1\frac{1}{8}$ to $1\frac{1}{4}$ feet (.3 to .4 m) long and weighs from $1\frac{1}{2}$ to $2\frac{1}{2}$ pounds (.7 to 1.1 kg). The fur is pinkish-brown, sprinkled with white above, and is a somewhat lighter brown below. The nose is yellowish, and there are dark patches above and below the eye. The short tail is tipped with white fur.

Like the related Black-tailed Prairie Dog, this prairie dog digs burrows, but with fewer interconnecting burrows. The White-tail is much less social than the Black-tail, and engages in very little touching and 'kissing' compared to the Black-tail. Their vocalizations, including alarm calls, are similar to those of the Black-tail.

The White-tailed Prairie Dog feeds on grasses and herbs that grow on the high sagebrush plains that comprise its habitat. White-tailed Prairie Dogs gain weight rapidly over the summer months in preparation for winter. True hibernators, they retire to their burrows around October and emerge in March. During the hibernation period they may awaken at times to feed on roots and in-ground seeds. They may also estivate for short times during the hottest parts of the summer.

Mating begins in March, and litters of four to five are born after a gestation period of about a month. The young hibernate with the parents the first year and disperse on emerging in the spring. Predators include badgers, coyotes, snakes and birds of prey.
prey, which are particularly feared.

In addition to the Black-tailed Prairie Dog (*Cynomys ludovicianus*), other similar species are Gunnison's Prairie Dog (*Cynomys gunnisoni*) and the Utah Prairie Dog (*Cynomys parvidens*), which are considered to be the same species as the White-tail by some biologists.

RACCOON
(Procyon lotor)

The Raccoon, one of the most frequently seen mammals in North America, is found in almost all of the United States, southern Canada and northern Mexico. Portions of the Rocky Mountain states are the only areas of the United States not included in the Raccoon's range. The Raccoon is 2 to 3⅛ feet (.6 to .9 m) long and weighs from about 10-50 pounds (4.5 to 23 kg). The fur is reddish-brown tinged with black above and is grayish on the underside. The long, bushy tail has alternating black and brownish-grayrings. A patch of black fur, lined in white, forms a black mask.

***Above:* Raccoons prowl by night**

The nocturnal Raccoon eats almost anything, including small mammals such as mice and voles, insects, and a wide variety of plant foods. It climbs trees, both to escape from predators and to get birds' eggs and nestlings, acorns and fruit. The Raccoon regularly hunts along stream banks and swims in the stream in pursuit of crayfish, fish, worms, frogs, insect nymphs, turtles and turtle eggs. The Raccoon commonly dips food in water, if near a stream or pond, not to wash it, but to soften it so that it can tear the food, discarding what it regards as unsuitable to eat. The Raccoon does well in areas of human population, regularly knocking over garbage cans and raiding food stores with its almost monkey-like hands. It can readily pry the tops off garbage cans, open cabinet and refrigerator doors and turn on water faucets. It does not hibernate but may sleep in its den during cold spells.

Raccoons are solitary except at breeding time which starts in December in the south and in February in the north. Typically, a male spends about a week denned up with a female and then leaves to find another one. Females build nests of leaves in dens that are usually hollow trees but can be in almost any protected place. Litters, born from April to May, usually number four or five, but can be as large as seven kits. When the kits are very young, the mother Raccoon carries them cat-fashion by the nape of the neck. When they are about four weeks old, they begin to follow the mother. Very protective, Raccoon mothers hiss at intruders, send the young up trees and fight viciously if cornered.

Although a Raccoon may occasionally raid a henhouse or take some corn or other produce from a farmer's field, it is not generally considered to be a serious agricultural pest. 'Coon' hunting is a popular sport, particularly in the South. Dogs are used to trail a Raccoon and drive it into a tree. The meat is considered to be excellent, with a taste somewhat like lamb. In colonial times coonskin caps and sleigh robes were common. Raccoons are still hunted for pelts, although much less so than in the 1920s when coonskin coats were popular. Kits make amusing pets, but become cranky and dangerous as they age. Similar species include the Coati (*Nasua nasua*) and Ringtail (*Bassariscus astutus*).

RABBIT, EUROPEAN; DOMESTIC RABBIT *(Oryctolagus cuniculus)*

The European Rabbit was introduced to North America first on Farallon Islands, California; San Juan Islands, Washington; and Middelm Islands, Alaska. Other groups were released in Pennsylvania, Illinois, Indiana, New Jersey, Maryland and Wisconsin. Today, the European Rabbit is found throughout the whole of North America, particularly in rural and suburban areas where food abounds.

The European Rabbit is from 1½ to 2 feet (.45 to .6 m) long and weighs from 3 to 5¼ pounds (1.3 to 1.6 kg). Variable in color, most European Rabbits are from brownish to grayish with white fur on the underside of the tail. However, some are almost white, while others are black.

Introduced in an effort to provide a game animal larger than the cottontails, the European Rabbit has thrived in North America. Active mostly at night, they are not as successfully hunted by man as are the Cottontails, which can be found in the open at dawn and dusk. They eat a wide variety of plant foods, including garden and crop plants, a circumstance that makes them a serious agricultural pest in many areas.

RABBIT, MARSH *(Sylvilagus palustris)*

The range of the Marsh Rabbit extends from Florida northward to extreme southern Alabama, southeastern Georgia and the eastern areas of South Carolina, North Carolina and the southeastern tip of Virginia.

The Marsh Rabbit measures from 1.2 to 1.5 feet (.36 to .45 m) long and weighs from 2½ to 3½ pounds (1.1 to 1.6 kg). The fur is dark brown above, white below. The nape of the neck is a dark cinnamon color, and there is reddish-brown fur on the feet. The ears are short compared to those of most other rabbits.

The Marsh Rabbit is somewhat unusual among rabbits in that it readily takes to water when threatened. Once in the water, it floats with only the eyes and nose exposed. With short legs, it cannot move as quickly on land as can most other rabbits.

***Above:* The Marsh Rabbit lives near water.**

The food of the Marsh Rabbit includes a great deal of marsh vegetation such as cane, rushes and other grasses of its bottomland, swamp and coastal waterways habitat. It also eats bulbs, leaves and twigs. Rushes are also used to provide protective cover.

The mating season begins in February, and several litters of four to five young are born every year in large, covered nests made of grass, leaves and rushes. The Marsh Rabbit, a widely hunted game animal, seldom invades human food-growing areas.

Similar species are the Swamp Rabbit (*Sylvilagus aquaticus*) and the Brush Rabbit (*Sylvilagus bachmani*).

RAT, BLACK; SHIP RAT; ROOF RAT; WHARF RAT *(Rattus rattus)*

Originating in central Asia, the Black Rat spread through Europe long before Columbus made his first voyage. It was well established in Europe by the time Europeans began to explore and colonize the New World. The Black Rat was first introduced into the western hemisphere in Central and South America by Spanish explorers, and the first Black Rats in North America came ashore from the ships that brought English colonists to Jamestown in 1607. Today, the Black Rat is found all along the Atlantic, Gulf and Pacific Coasts. Inland, it is most abundant in the South and the southwestern United States. Its range was once much more extensive, but it has been pushed out of much of its former range by the more aggressive Norway Rat.

The Black Rat is from 1 to 1½ feet (.3 to .45 m) long and weighs from 4 to 12 ounces (113 to 342 g). The fur is brown to black above and a light gray below. The scantily-haired, scaly tail is longer than half the total body length.

Found primarily in and around seaports and city buildings, the Black Rat is also found in fields and woods, but much less so than the Norway Rat. The Black Rat is totally omnivorous, but has a particular liking for grain. Contaminating what it does not eat with its droppings, it is a particular problem around docks and warehouses. Although it is not as aggressive as the Norway Rat, it is much more agile. It climbs the sides of buildings and crosses from one building to another on telephone and electric wires. It is commonly found in the roofs of buildings in the South, and it also nests in trees. Breeding thoughout the year, a female can produce several litters of two to eight young.

The Black Rat carries a number of diseases including bubonic plague and typhus. Eating rat-contaminated food can possibly result in a parasitic worm infestation. Its chief predators are snakes, owls, domestic dogs and cats.

RAT, GIANT KANGAROO *(Dipodomys ingens)*

Limited in range to the western San Joaquin Valley of California, this Kangaroo Rat is distinguished by being the largest of the kangaroo rats. The Giant Kangaroo Rat is from 1 to 1⅛ feet (.3 to .35 m) in length, of which some 6 to 7¾ inches (15 to 20 cm) is tail. It weighs from 4½ to 6¼ ounces (127 to 177 gm).

Like most kangaroo rats, the Giant Kangaroo Rat eats seeds, but it also

Above: **The Norway Rat invades homes.**

eats the green parts of plants. Seeds are eaten on the spot and carried in cheek pouches to burrows for storage. When disturbed, the Giant Kangaroo Rat can make leaps of nine feet (2.7 m) or more. The feet are also used to drum the ground as a danger signal. Tapping a burrow sometimes induces the kangaroo rats inside to drum in response. Breeding takes place from January to May. One or two litters of three to five young are born each year.

Similar species include the Heerman (*Dipodomys heermanni*), Pacific and Santa Cruz Kangaroo Rats.

RAT, MARSH RICE; RICE RAT
(Oryzomys palustris)

The only rice rat with an extensive range in the United States, the Marsh Rice Rat is found primarily in the southeastern United States from eastern Texas and Oklahoma, eastward to the Carolina coast, including Florida, northward along the Atlantic coast to eastern Maryland, Delaware, southeastern Pennsylvania and southern New Jersey. It is also found in Kentucky, southern Indiana and Illinois, southeastern Missouri and a small area of southwestern Missouri.

The Marsh Rice Rat is adapted for life around marshes and grassy or sedge-filled areas. A strong swimmer, both on the surface and under water, this rice rat has a varied diet including tender parts of aquatic plants, crabs, snails, insects, berries and other fruits. It also eats the underground fungus *Endogone*.

Mainly nocturnal, the Marsh Rice Rat constructs runways on land and builds nests in debris above the high water mark. Breeding throughout the year, a female has three to four litters of one to seven young annually. Females are sexually mature at about age four weeks.

Although it may occasionally cause minor damage in a rice field, it seldom causes serious losses.

RAT, MOUNTAIN PACK
see Woodrat, Bushy-tailed

RAT, NORWAY; BROWN RAT; WATER RAT; SEWER RAT
(Rattus norvegicus)

Originating in central Asia, the Norway Rat spread across Europe in the 16th, 17th and 18th centuries. It is believed to have been introduced into North America in grain shipments sent for the Hessian troops hired by the British to fight in the American Revolution. Reintroduced again and again on ships from all parts of Europe, the Norway Rat is found today in all parts of the United States and in southern Canada. The name Norway Rat is a misnomer, stemming from the once-held belief that the animal was introduced into North America from a Norwegian ship.

The Norway Rat is from 1 to 1½ feet (.3 to .45 m) long and weighs from 6 to 17 ounces (170-482 g). Often heard reports of Norway Rats as large as house cats or medium-sized dogs are exaggerations. The fur is brownish-gray above and gray below. The long snaked tail is scaly, and comprises somewhat less than half the total body length.

The totally omnivorous Norway Rat will eat practically anything, including material generally not thought of as food. It seems to have a particular liking for stored grains which it will consume in great quantities if it can get into storage areas, and it contaminates what it does not eat with its droppings. It is also fond of chicken eggs and young chicks. In houses and buildings they gnaw wallboard, woods, plaster, furniture and electric wires, sometimes causing fires.

The Norway Rat digs interconnecting tunnels, usually about 1½ feet

(.45 m) deep and 6 feet (1.8 m) long, with chambers for nesting and feeding. It digs with teeth and forefeet. Able to adapt to a variety of habitats, the Norway Rat lives in fields, farms, all kinds of buildings and on ships. They can gnaw through brick walls to get into a building.

Breeding at any time of the year, the Norway Rat has an enormous reproductive potential. Females, which sometimes mate again only hours after delivering a litter, are capable of bearing as many as 12 litters a year. Many factors, however, such as food supply, usually keep the total far below that number. However, the usual production of 5 litters of seven to ten young still exceeds the rate of most other rodents. The young, born blind and without fur, open their eyes at two weeks and are weaned at three weeks. Females are ready to breed at age three months.

Mass migrations may occur if local food supplies become unusually scarce. These mass migrations are possibly the basis of the Pied Piper legend.

In addition to destroying food and property, these rats occasionally bite and severely injure sleeping people, particularly children and infants. They also carry diseases such as typhus, bubonic plague and tularemia. Albino strains of Norway Rats are widely used as laboratory animals. The Black Rat (*Rattus rattus*) is a similar species.

RAT, ORD'S KANGAROO
(Dipodomys ordii)

The range of Ord's Kangaroo Rat extends from northern Mexico to the southern tip of Texas, northwestern Texas, northward through the Rocky Mountain states including all of New Mexico, Colorado and Utah, most of Nevada, Arizona and Wyoming, eastern Montana, southern Idaho, eastern Oregon, a small area of south central Washington and the northeastern fringe of California. It is also found in the western parts of the tier of states from Oklahoma to South Dakota, and in southwestern Saskatchewan and southeastern Alberta.

***Above:* An Ord's Kangaroo Rat.**

More closely related to ground squirrels and pocket gophers than to rats, Ord's Kangaroo Rat is from $8\frac{1}{8}$ to 11 inches (21 to 28 cm) long, of which $3\frac{3}{4}$ to $6\frac{3}{8}$ inches (9.5 to 16 cm) is the tail. Its weight ranges from $1\frac{3}{4}$ to $3\frac{5}{8}$ ounces (50 to 102 g). The fur is buff above and white below, and there may be white spots at the base of the ears.

Found in arid, sandy areas, Ord's Kangaroo Rat forages on the ground for seeds, mostly those of sunflower, mesquite, tumbleweed, Russian thistle and sandbur. A nocturnal burrowing animal, it is active throughout the year in Texas. In the northern parts of the range it spends most of the winter months in its burrows, subsisting on stored food. The burrows contain chambers for nesting, food storage and waste disposal.

Normally, Ord's Kangaroo Rat moves about on the ground in a rapid scurry. However, if it is disturbed, it can make hops of from 6 to 8 feet (1.8 to 2.4 m). The large, spring-like hind feet, which give the animal its name, are also used to kick sand in the face of a predator and to thump the ground as a danger signal to other kangaroo rats. During the leap, the tail helps to maintain balance.

This and other kangaroo rats eat some food on the spot and carry some in their cheek pouches for transport to the burrow for storage. They obtain their water needs from the seeds they eat, an adaptation for their dry habitat. Ord's Kangaroo Rat also takes sandbaths to keep its fur from getting matted.

One or two litters of two to five young are born a year. Litters are usually born in May and June.

Ord's Kangaroo Rat has little if any impact on the economic activity of man. They are major food for carnivores, snakes and birds of prey including badgers, skunks, foxes, weasels, coyotes, rattlesnakes and owls.

Similar species include other kangaroo rats, particularly the Panamint Kangaroo Rat (*Dipodomys panamintinus*) and the Chisel-toothed Kangaroo Rat (*Dipodomys microps*).

SEA LION, CALIFORNIA
(Zalophus californianus)

The range of the California Sea Lion extends from Vancouver, along the Pacific Coast southward to Baja California, and into the Gulf of California.

The males, much larger than the females, are from $6\frac{1}{2}$ to $8\frac{1}{4}$ feet (2 to 2.5 m) long and weigh from 400 to 650 pounds (182 to 295 kg). The females

***Above:* California Sea Lions at rest.**

are from 5 to 6½ feet (1.5 to 2 m) in length and weigh from 100 to 225 pounds (45 to 102 kg). The fur varies from buff to brown, but appears to be a shiny black when the animal is wet. The males have a distinctive high forehead.

The California Sea Lion is the trained seal of circuses, aquariums and animal shows. As playful in the wild as it is in captivity, the California Sea Lion frequently throws objects, catching them on its nose, and cavorts with other sea lions. Highly gregarious, these sea lions hunt and play in separate male and female groups, coming together only in the breeding season.

The California Sea Lion spends much of the day sleeping on shore. It hunts at night, and its diet includes octopus and abalone, in addition to the fish which are its main food. The fastest of the sea-going carnivores, it can swim gracefully at speeds in excess of 25 mph (40 kph). Capable of remaining submerged for as long as 20 minutes, it locates prey and finds it with a 'sonar' method of emitting sounds and interpreting the echoes. Like other eared seals, the California Sea Lion can move fairly well on land using its folded flippers as 'feet.'

The breeding season begins in June, when males and females come ashore on California's Channel Islands, and ends in July. The males establish territories which they defend to the accompaniment of loud, honk-like barks. Unlike the Northern Fur Seal, the male California Sea Lion defends territories, not harems, and mates with any female in his territory. A male does not mate until it is large enough to defend a territory.

Females give birth to single, blue-eyed pups weighing about 36 pounds (16 kg) at birth, and mate shortly afterwards. Females vocalize with an undulating wail that is a call for the pup, and with growls and barks of annoyance directed at other females. Pups recognize the mother's vocalizations and respond with lamb-like bleats. The pup stays with the mother for about a year, nursing, eating fish and learning to swim.

Once hunted for its blubber and meat, which was used for dog food, the California Sea Lion is today fully protected by law. Many are taken by killer whales and several species of sharks.

Similar species include the Northern Fur Seal (*Callorhinus ursinus*) and the Northern Sea Lion (*Eumetopias jubatus*).

SEA LION, NORTHERN: see Sea Lion, California; Seal, Northern Fur

SEA WOLF: see Whale, Killer

SEAL, ALASKA FUR: see Seal, Northern Fur

SEAL, GRAY; see Seal, Northern Elephant

SEAL, HAIR: see Seal, Harbor

SEAL, HARBOR; LEOPARD SEAL; HAIR SEAL *(Phoca vitulina)*

The Harbor Seal is found on the Arctic Ocean coast, south into Hudson Bay and in the freshwater Seal Lakes, south along the Atlantic Coast to the Carolinas, along the entire coast of Alaska, the Pacific coast of Canada, and south along the coast into California.

Unlike most other seals and sea lions, male and female Harbor Seals are about the same size. The Harbor Seal ranges in length from about 4 to 5½ feet (1.2 to 1.7 m) and weighs from about 250 to 300 pounds (113-136 kg). Harbor Seals are quite variable in color, ranging from cream to dark brown above and a creamy white with spots below. However, yellowish-gray or brownish fur above with spots is the most prevalent coloration.

The Harbor Seal is an example of an earless seal. Often called 'true seals,' these seals have stiff fur as adults and no external ears. While the hind flippers are highly efficient swimming tools, they are permanently turned backward and cannot be folded for movement on land. Consequently, the Harbor Seal and other earless seals cannot move well on land, and are propelled only by muscular contraction of the entire body. Whenever possible, they take advantage of inclines for rolling or sliding down the ice.

Highly gregarious, Harbor Seals can often be seen in groups of 500 or more individuals resting on beaches and rocky shores. Very strong, graceful swimmers, they can dive as deep as 300 feet (97 m) and stay underwater for almost 30 minutes.

The Harbor Seal's diet is mostly fish such as herring, cod, mackerel, flounder and salmon, supplemented by squid, clam, octopus, shrimp, crabs and crayfish. The feeding activity is often keyed to the tide. As the tide comes in, they go up the river with the tide, going ashore to sleep when the tide is low and taking to the river again at the next high tide. In the spring they may follow the movements of fish several hundred miles upstream, and return to the sea in the fall.

Pups are born from March to August. Although males usually mate with several females, they do not form harems or maintain mating territories as do many other seals and sea lions. Pups usually enter the water for the first time when they are a few days old.

Harbor Seals sometimes tear up the nets of commercial fishermen to get at the fish inside. The flesh of the Harbor Seal is used as food in some parts of its range, but the fur has little, if any, commercial value.

The Spotted Seal (*Phoca largha*) is a closely related species and the Ribbon Seal (*Phoca fasciata*) is similar.

Above: **Harbor, Leopard or Hair Seal.**

SEAL, HOODED:
see Seal, Northern Elephant

SEAL, LEOPARD:
see Seal, Harbor

SEAL, NORTHERN ELEPHANT
(Mirounga angustirostris)

The Northern Elephant Seal is found along the Pacific coast from the Gulf of Alaska to Baja California.

The largest of the sea-going carnivores, the female Northern Elephant Seal is from 9¾ to 11½ feet (2.9 to 3.5 m) long and weighs up to about 2000 pounds (908 kg) and the male is from 14¾ to 21⅓ feet (4.5 to 6.5 m) long and weighs up to 7800 pounds (3500 kg). The fur is brown or gray in the upper regions and a bit lighter below. The male has a large snout that droops over the muzzle and is inflated during the mating season. Like all earless or true seals, the flippers of the Northern Elephant Seal are not well adapted for movement on land.

From August to December, the Northern Elephant Seal stays out at sea, feeding on fish such as ratfish, hagfish, small sharks and other deepwater marine life such as squid. A strong swimmer, the Northern Elephant Seal can dive to depths of 200 feet (61 m) and remain submerged for as long as 40 minutes.

Starting in December, the seals start to return to islands off the California coast for breeding. Bulls arrive first, and immediately begin to fight for harem territories. The fighting starts when the bulls threaten each other by inflating their huge snouts and bellowing and roaring at each other. The inflated snouts act as resonating chambers which make their vocalizations audible for more than a mile. Northern Elephant Seals fight by slamming into each other with their massive necks and biting whenever they can. The territorial disputes are generally settled by the time the females arrive, and the bulls concentrate on building their harems.

Females begin to come ashore in late December and early January. Soon after arrival, the female delivers a single, black-furred, 65 pound (29.5 kg) pup. After a month of nursing on the mother's rich, 54.5% fat-content milk, the pup grows to a weight of about 400 pounds (182 kg). When the pup is weaned, the mother abandons it and again becomes available for mating. Mating takes place from January to March, and during that time many pups are crushed by bulls lumbering over them to get at females. During the mating seasons, the bulls do not eat, but females continue to take feeding swims. Females

Above: **A sleeping male Northern Elephant Seal, the largest North American seal.**

begin to molt in late May, and bulls begin their molt in July and August. Shortly after the molt is complete, the seals return to the sea. The weaned pups live on accumulated fat and teach themselves how to swim until May when they, too, take to the sea.

Widely hunted in the 19th century by whalers for their blubber, the Northern Elephant Seal was believed to be extinct or near extinction when a small colony was found on Guadalupe Island off Baja California in 1892. The animals were placed under protection, and today they number about 65,000. The increase in population has forced some elephant seals off the islands onto mainland beaches. There is a rookery near Santa Cruz, California.

Similar species include the Hooded Seal (*Cystophora cristata*) and the Gray Seal (*Halichoerus grypus*).

SEAL, NORTHERN FUR; ALASKA FUR SEAL *(Callorhinus ursinus)*

Like most seals and sea lions, the Northern Fur Seal spends most of its life at sea, coming ashore to mate, give birth and rest. They come ashore at points on the Pacific Coast and Arctic Ocean coast ranging from Point Barrow, Alaska southward to San Diego, California, depending on the season.

As is the case with most of these ocean-going carnivores, the bulls are considerably larger than the females. Bulls are from 6¼ to 7¼ feet (1.9 to 2.2 m) long and weigh from about 235 to 600 pounds (147 to 272 kg). The fur of the bull is black on the upper parts of the body, except for a grayish cast over the massive shoulders, and reddish below. Bulls have enlarged, bulging necks. Females are from 3⅔ to 4⅔ feet (1.1 to 1.4 m) long and weigh from about 85 to 120 pounds (38 to 54 kg). The fur of the females is gray over all regions of the upper body, and somewhat reddish below.

Belonging to the eared seal group of ocean-going carnivores, the Northern Fur Seal has small but observable external ears. Although highly adapted to life in the oceans, the eared seals can move on land with greater facility than can the earless seals. The hind flippers rotate under the body, and the fore flippers turn out at the wrist to provide somewhat crude but serviceable 'feet' with which they can walk or move more rapidly in a lumpy gallop. When swimming, the fore limbs are used for propulsion and the rear flippers are held together to work as a rudder.

The Northern Fur Seal probably spends more time and travels greater distances in the ocean than do other seals and sea lions. Returning to land only for breeding, the Northern Fur Seal spends most of the year far out at sea, hunting for its major foods of fish and sea birds. It rests in the ocean by floating on its back, bending the hind flippers over the belly and holding them there with a fore flipper. The seal does most of its feeding at night when fish come closer to the surface. It is capable of diving as deep as 450 feet (137 m) and staying submerged for several minutes at a time. It swallows smaller fish while submerged, and comes to the surface to bite larger prey into smaller pieces.

Northern Fur Seals begin to arrive in the spring on land at rookeries to give birth and mate. The largest and most extensive rookery is on the Pribilof Islands in the Bering Sea. Other rookeries are on San Miguel Island off the California coast and two small islands – Commander and Robben – near the USSR. In most years, more than a million seals gather in the 30 square mile Pribilof Islands. The first arrivals are older bulls who come ashore around May and early June. They immediately fight viciously to establish reign over the best spots, those near the water where the females come ashore. The females start to arrive around the middle of June and continue to come in until around mid-July. As the females arrive, the bulls gather them into harems that can number from a few, to as many as 100 or so, depending on the size of the bull and his ability to outbluff and outfight other bulls. In general, the larger the bull, the larger the harem is likely to be. During the two-month breeding season, the bulls do not eat. They spend all their time mating and defending their harems.

The females give birth to a single, glossy black, 10 to 12 pound (4.5 to 5.4 kg) pup, the result of the previous year's mating, a few days after coming ashore. Females mate three days to a week after giving birth, usually with the bull who reigns over the territory in which she gave birth, unless she has been 'stolen' by another bull in the meantime. Implantation of the fertilized egg is delayed so that birth will coincide with the next year's breeding season.

The female remains with the pup for about a week, and then goes out to feed, returning at intervals to nurse the pup with her very rich milk that is about 50% fat. Cow's milk, in comparison, is about 3.5% fat. Pups stay with the mother and those that survive the crowded, hectic conditions of the rookery, leave when the seals depart for winter feeding areas.

The final phase of the breeding season occurs in August, when females who have not yet bred come ashore. These females usually mate with bulls too young and small to maintain harems and territories. The seals begin to leave the rookery, shortly after the arrival of the young females, to migrate to winter feeding areas. The bulls, thin, weak and battle-scarred, leave in August to spend the winter south of the Aleutian Islands, the Gulf of Alaska or areas off the Pacific Coast of the USSR.

Females and juveniles remain on the rookery until November, when they start to make their separate ways to winter areas. Females and juveniles migrate much greater distances than do bulls. Some swim more than 3000 miles to southern California.

One of the most important commercially hunted animals in the world, Northern Fur Seals are killed for their fur and their blubber, which is rendered into oil. The flesh is sold for meat, primarily mink food, and the bones are ground up into a meal. Overhunting had brought the Northern Fur Seal near extinction by the mid-19th

century. Since then international treaties have afforded some protection. Hunting quotas are set, and the populations are carefully managed. Killing of females is forbidden; allowable kills are limited to males in bachelor herds. Similar species include the Northern Sea Lion (*Eumetopias jubatus*) and the California Sea Lion (*Zalophus californianus*).

SHEEP, BIGHORN; MOUTAIN SHEEP; ROCKY MOUNTAIN BIGHORN SHEEP *(Ovis canadensis)*

In many parts of the Rocky Mountains, the Bighorn Sheep is more often heard than seen. Generally found in areas away from human habitation, they are rarely seen by the casual passerby. However, when they engage in their horn-butting battles, the resounding crack that is made as they crash together can be heard for a mile or more, depending on conditions.

The range of the Bighorn Sheep is from Mexico north through western California (the Sierra Nevada), Nevada and the Rocky Mountain states to British Columbia and Alberta. They spend much of their time in meadows and foothills close to rocky cliffs. They are also found in desert regions such as Death Valley.

One of the larger of the sheeps, the Bighorn Ram is 3-3½ feet (.9-1.05 m) high at the shoulder, 5 to 6 feet (1.5-1.9 m) long, and weighs from 125 to 320 pounds (57-145 kg). The smaller female is 2½ to 3 feet (.75-1.0 m) at the shoulder, 4 to 5 feet long (1.2-1.5 m) and weighs from 75 to 200 pounds (34-90 kg).

The most notable physical feature is the pair of massive curved horns on the ram. The horns curve backwards, downward and then forward past the cheeks, forming a huge C. Horns that measure more than 4 feet (1.2 m) along the length of the curve have been observed. The full C shape of the horn is called a 'curl.' Some older animals, ages 7 to 8 and older, have curls that curve around until the tips are level with the base of the horns. Ewes have short, slender horns that curve slightly. The hooves of the Bighorn are hard along the outer rims and spongy in the center, giving the animal good traction. In addition to being a good climber and jumper, it is also a strong swimmer.

The color varies from a pale tan to dark brown. The darker color is more likely to be seen in the northern mountainous part of its range, and the lighter color is more prevalent in the desert regions.

Bighorn Sheep eat grasses, sedges and woody plants. In the summer they migrate to higher slopes, and return to the valleys in the winter. Active through most of the daylight hours,

***Above:* A Bighorn Ram high in the Rocky Mountains of Banff National Park.**

they feed in the morning, afternoon and early evening. When they are not feeding they lie down and chew their cuds. A Bighorn Sheep tends to use the same spot for bedding down at night, returning to it every night for years and years.

Bighorn Sheep live in small herds of about ten animals that include ewes, lambs, yearlings and two-year-olds. In winter, the size of the herds grows to about 100, when the rams rejoin the group. The herd leader is usually an old ewe. The rams leave the herd in the spring, moving to the summer range higher in the mountains. Many Bighorns have been killed by trophy hunters and their numbers further depleted by stock-grazing and disease.

The start of the mating season is heralded by the rams' butting fights and the mating season runs from August to early January, depending on the latitude. As the mating period approaches a peak, the butting fights become almost continuous. The rams stand about 20 feet apart, lower their heads and charge, coming together with a loud crack. In spite of the intensity of the contests, the animals are seldom injured. Status is determined by horn size and fights almost always occur between males with similarly-sized horns. Males will follow the nearest ewe in heat, and fights occur whenever two or more rams follow the same ewe. Mating can occur any time during the August to January mating season.

Well-developed, wooly lambs are born after a gestation period of 180 days. Births are usually single; twins occur occasionally. Ewes hide the lambs for a week, after which the lambs follow their mother. Ewes are sexually mature at about two-and-a-half years of age. Rams stay with the small ewe-lamb until they are two, when they join groups of rams.

Bighorn Sheep have a life span of about 15 years. They are preyed on by wolves, coyotes and cats such as bobcats and mountain lions.

Similar species include the Dall Sheep (*Ovis dalli*) and the Mountain Goat (*Oreamnos americanus*).

SHREW, SHORT-TAILED
(Blarina brevicauda)

One of the most abundant mammals in North America, the Short-tailed Shrew ranges from southeastern Canada, south to the northeastern United States, westward to Minnesota, Iowa, the Dakotas, and Nebraska, and southward into Virginia, West Virginia, Kentucky, and along the Allegheny Mountains into Georgia and Alabama.

Like all shrews, the Short-tailed Shrew is small compared to other mammals, but is the largest shrew in North America. It is from 3¾ to 5 inches (9.5-13 cm) long and weighs from ½ to 1 ounce (15 to 30 g). It is brownish, and, as its name suggests, it has a short tail that measures from ¾ to 1 inch (2-2.5 cm) in length. The tails of most other shrews are four to six times longer.

The Short-tailed Shrew uses its snout and forefeet to dig underground runways in a search for prey. When it bites its prey, it paralyzes it with a poison in its saliva, and the prey is dragged to the shrew's underground nest to be eaten. Eating more than its body weight in a day, feeding frequently to satisfy the needs of a very high rate of metabolism, the Short-tailed Shrew's diet includes earthworms, insects, snails, centipedes, small mice and other shrews, plus underground fungi. They are preyed on by owls, snakes and other mammals, but many are killed and not eaten, apparently because of an odor that is repellant to some predators.

The Short-tailed Shrew will usually fight viciously any animal it perceives to be an intruder, particularly other Short-tailed Shrews. Fights between Short-tailed Shrews are fierce and intense, but usually end without serious injury when one lies on its back in a submissive position. Males mark off territories in their burrows with secretions from glands on the belly and hip. During the mating season, these markings prevent female-hunting males from entering the burrows of other males. Complex courtship rituals usually prevent fights between sexes.

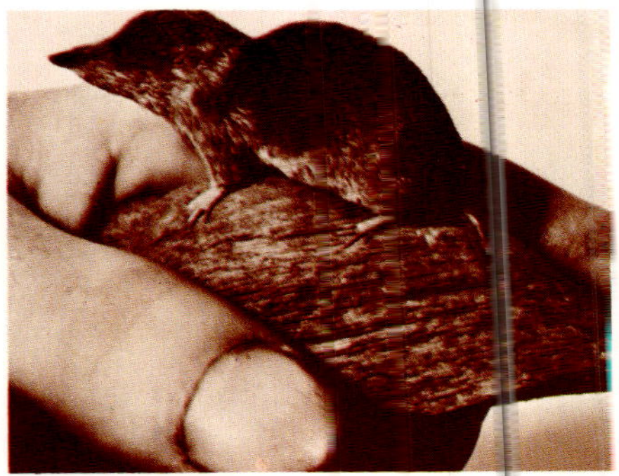

Above: A Short-tailed Shrew, the largest North American Shrew.

Litters of five to six young are born in the spring or summer in aboveground nests of leaves or grass built under stump or log. When the family group is alarmed, each young takes the base of the tail of a litter mate in its mouth, while one takes the tail of the mother in its mouth. The resulting chain of shrews can then be dragged to safety by the mother. Almost all species of shrews exhibit this behavior.

Closely related species are the Swamp Short-tailed Shrew (*Blarina telmalestes*) and the Southern Short-tailed Shrew (*Blarina carolinensis*).

SHREW, WATER; NORTHERN WATER SHREW
(Sorex palustris)

This shrew is found in most of Canada except Newfoundland and the northern areas, and southern Saskatchewan and Alberta. The range extends southward into New England, New York, Pennsylvania, West Virginia, the mountainous area where Kentucky, Tennessee, and Virginia meet, northern Michigan, Minnesota and Wisconsin. It is also found in Washington, Idaho, eastern Montana, Wyoming, and Colorado, most of Utah, eastern Nevada, north central New Mexico, north central California and a small area in Arizona

The Water Shrew is from about 4 to 6½ inches (10-16.5 cm) long and weighs about ⅜ to ⅝ ounces (11-18 g). Well adapted to life in and around water, the Water Shrew has dark brown, velvety, water-resistant fur. The hind feet have fringes of stiff hair which increase its buoyancy and improve its swimming ability.

The Water Shrew swims and dives into small streams, searching for a variety of aquatic organisms such as the immature nymphal forms of mayflys and stoneflys. It also eats terrestrial insects and other invertebrates. It commonly dives to the bottom, and when it stops its swimming motions, air trapped in its fur pops it to the surface. The fringe on the hind feet increases the surface area to the point where the Water Shrew can actually run on the surface of the water in short, quick bursts. In the water, it is preyed on by bass, pickerels, pike and large trout. When the Water Shrew goes ashore after swimming, it dries itself with the stiff hind leg hairs that act as a comb. On land, Water Shrews are often eaten by weasels.

Litters of four to eight young are born from February through June. It is thought that females, which are ready to breed at age three months, have more than one litter a year.

SKUNK, EASTERN SPOTTED; SPOTTED SKUNK
(Spilogale putorius)

The range of the Spotted Skunk extends from Mexico northward to include most of the United States except Montana, western Washington and northern North Dakota, New England, the Middle Atlantic states, the Midwestern states east of the Mississippi and the coastal areas of Virginia, North Carolina, South Carolina and Georgia.

Smaller than other skunks, the Spotted Skunk is from about 1.1 to 1.8 feet (.3 to .52 m) long and weighs from 1.7 to 2¼ pounds (.7 to 1 kg). It is black with horizontal white stripes on its neck and shoulder and it has irregular stripes and elongated white spots on the sides.

The Spotted Skunk moves with greater speed, agility and grace than other skunks. It is a good climber, often taking to trees to avoid predators and to look for food. More carnivorous than other skunks, the Spotted Skunk's diet consists mostly of small mammals, particularly mice, but it will also eat insects, insect grubs and a variety of plant foods including grapes and corn.

Above: **The Striped Skunk is an omnivore.**

The Spotted Skunk's distinctive coloration advertises its ability to spray a foul-smelling musk from its anal glands. Once a potential predator has received a burst of the pungent spray, it is unlikely to try to take a Spotted Skunk again. If the predator continues to press its attack, the Spotted Skunk turns its back to the predator. Then, standing on its forefeet, it raises its tail, spreads the hind feet out, and sprays over its head. The spray can carry for a distance of 12 feet (3.6 m) or more. Most predators who have learned by experience will back off when they see the Spotted Skunk go into the preliminaries to spraying.

The Spotted Skunk dens in rockpiles, burrows and under buildings. While other skunks den alone, the Spotted Skunk often dens up in groups, particularly in the winter. The winter inactivity of the Spotted Skunk is not a true hibernation.

Mating takes place in late winter. Litters of four to five are born the following spring. By summer they have adult coloration and follow the mother. The Great Horned Owl sometimes takes young Spotted Skunks by striking from the air before the mother skunk can spray.

SKUNK, STRIPED; COMMON SKUNK
(Mephitis mephitis)

One of the more widely distributed mammals in North America, the Striped Skunk is found in the southern half of Canada southward through all the continental United States except the extreme southern tip of Florida and into northern Mexico. Members of the weasel family, both the Striped and Spotted Skunk are found only in North and South America.

The Striped Skunk is from 1.7 to 2.6 feet (.5 to .8 m) long and weighs from 6 to 14 pounds (2.7 to 6.3 kg). Most are black with two broad white stripes running down the back to the middle of the forehead. There is a broad patch of white fur on the back of the neck. The bushy black tail is usually tipped with white. Some individuals are grayish to almost white.

The omnivorous skunk has a varied diet that includes insects, small mammals, birds' eggs, amphibians and various plant foods. Primarily nocturnal, they walk in a shuffling, waddling, bouncy gait, stopping often to dig for insect grubs and other food.

The best-known adaptation of the striped and other skunks is its ability to spray a strong-smelling oil from its anal glands. Other animals soon learn to recognize the rather conspicuously colored skunk and to avoid it if possible. Only the great horned owl, which can swoop and pick up a young skunk before the mother can spray, is a serious predator. A skunk will often spray in the direction of what it perceives as a threat. A mother skunk with young is quite likely to spray if disturbed. The skunk goes through a behavior that serves to warn the intruder of the intention to spray. First it snarls, then it stamps its feet, raises its hind legs, clicks its teeth, arches its tail over its back and then turns its rump toward the intruder and sprays. One jet or burst of the anal gland oil – about a fifth of a tablespoon – spreads out some ten to fifteen feet, and the visible mist can diffuse some 50 feet. The smell can carry for a mile or more, although the mist cannot be seen at that distance. If the fluid gets into the eyes, intense pain results. Contrary to a widely repeated myth, a skunk can spray if held by the tail.

Striped Skunks mate in late winter. Litters of four to seven young are born in May. By late June or early July they are weaned and follow the mother in single file on hunting walks. They can spray when they are six to seven weeks old.

Skunks eat constantly in the fall to store fat for the winter. In very cold weather they may hole up in a den, usually an abandoned burrow of another animal, and become dormant. However, their winter sleep is not a true hibernation. If necessary, a Striped Skunk will dig its own den, or it may den up in a hollow log or under a house or building. Striped Skunks denning in crawl spaces of houses are a frequently encountered problem.

Striped Skunks live in a variety of habitats including woods, plains and suburban areas. Although seldom seen in the daytime, the smell of skunks' spray is quite common in many populated areas in the evening or early morning. Many are killed on roads and highways, and the death of a skunk is often broadcast for miles as it gives off spray in its death agony. Dogs, particularly puppies and young dogs who have not learned to avoid skunks, are frequent spraying victims.

A Striped Skunk may, on very rare occasions, take a chick or an egg from a poultry house. However, it eats large numbers of harmful rodents and insects. Pelts, although not very valuable, are used to make fur products. Descented Striped Skunks are sometimes kept as pets. It is the chief carrier of rabies in the United States.

Above: **Like adults, young Striped or Common Skunks waddle or shuffle when they walk.**

SQUIRREL, FOX; EASTERN FOX SQUIRREL
(Sciurus niger)

The Fox Squirrel is found in all the United States east of the Mississippi River except New England, New York, New Jersey, northeastern Pennsylvania and extreme northern Wisconsin. West of the Mississippi it is found in the tier of states from Louisiana to Minnesota, and in almost all of the tier of states from Texas to North Dakota. It is also found in northern New Mexico along the east Texas border.

The largest of the tree squirrels, the Fox Squirrel is from about $1\frac{1}{2}$ to $2\frac{1}{3}$ feet long (.45 to .7 m) and weighs from 18 to 38 ounces (510 to 1075 gm). Three color phases are known to exist. In the Northeast, the fur is gray above and yellowish below; in the South the fox squirrel is black but may have a white blaze on the tail and face; in the West, this tree squirrel is a rusty color. Like all tree squirrels it has a large, bushy tail.

The Fox Squirrel is most likely to be seen in and around nut trees in the early morning and late afternoon. Its major foods are hickory nuts and acorns, but it also eats other nuts, seeds, fruits, berries and corn. It spends much of its time sitting in trees, eating or sunbathing. It also spends some time on the ground burying seeds. Like other tree squirrels, it does not remember where it buries a nut, but it can find them with a keen sense of smell even through several inches of snow.

Right: **The Fox Squirrel eats mainly acorns.**

In the summer, the Fox Squirrel may build a nest of leaves in a tree. Although it does not hibernate, it spends much of the winter in a tree hole, usually in a family group of several squirrels.

The winter mating season is heralded by males chasing females before the latter come into heat. Litters of two to four young are born after a gestation period of 44 days. Litters are usually born in February or March, but some are also produced in June or July and occasionally in late August or early September. From the age of two years onward, females usually have two litters a year.

The Fox Squirrel causes some damage to crops, particularly corn. It is a game animal in some rural areas, and is an attractive and interesting animal to watch in city parks. It has been introduced into some west-coast cities, notably San Francisco and Seattle.

Similar species include the Red Squirrel (*Tamiasciurus hudsonicus*), Eastern Gray Squirrel (*Sciurus carolinensis*, and Eastern Red Squirrel.

SQUIRREL, GRAY; EASTERN GRAY SQUIRREL
(Sciurus carolinensis)

In all probability, the most frequently seen wild mammal in the eastern United States is the Gray Squirrel. It is found in all states east of the Mississippi River except extreme northern Maine and a small area in western Florida. It is also found in the tier of states from Louisiana to Minnesota, westward to the eastern areas of the tier of states from Texas to North Dakota. In Canada it is found in the southern Maritime Provinces, southern Ontario, Quebec, Manitoba and Saskatchewan. The Gray Squirrel thrives in a variety of deciduous tree habitats from forests to suburban areas to city parks and streets.

The Gray Squirrel is from 1⅜ to 1⅔ feet (.4 to .5 m) long and weighs from 15 to 25 ounces (425 to 710 g). The fur is a dark gray color above and a paler gray on the underside. The flattened tail is covered with bushy gray silver-tipped fur. Black and albino phases exist in some parts of the range

Moving adeptly both on the ground and in trees, the Gray Squirrel is most active in the morning and early evening. Most of its diet consists of nuts, such as walnuts, acorns, beechnuts, hickory nuts and chestnuts, as well as various soft fruits, maple tree seeds and other seeds and corn. Sometimes it eats nuts where it finds them, but it usually buries the nut at a spot other than where it found it. Each nut is buried individually in a hole the squirrel digs with its forefeet, and is tamped down with fore and hind feet and nose. It does not remember where the nuts are buried, but can sniff them out, even under a foot or so of snow. The nut-burying activity of the Gray Squirrel serves to plant trees and spread them to new areas.

While the Gray Squirrel does not hibernate, it does hole up in a den in particularly cold weather. Dens and nests are established by both sexes, usually in trees, either in cavities such as woodpecker holes or in leaf nests in thick-limbed older trees, living or dead. Winter nests are much sturdier and tighter than summer nests. The summer nests are not necessarily in the same location as the winter nest. Loosely-constructed temporary nests are sometimes constructed near corn fields.

The Gray Squirrel moves smoothly through the trees by leaping from branch to branch. The bushy tail provides some lift during the leap, slowing descent in case of a fall. The tail also serves as an umbrella in the rain, a rudder for swimming and an effective blanket in the winter. They vocalize in a chattering bark.

Gray Squirrels mate in mid-winter. Litters of two to three young are born in the spring in leaf nests or tree-hole

***Below:* Red or Pine Squirrel.**

dens; there may be a second litter in the summer. Females often move their young between dens and leaf nests.

The Gray Squirrel is a widely hunted game animal. The meat, considered to be excellent, is a diet staple for people in some parts of the country. The Gray Squirrel occasionally causes crop damage, particularly to corn, and it may nest in the eaves of houses, causing some structural damage.

Similar species include the Eastern Fox Squirrel (*Sciurus niger*) and Red Squirrel (*Tamiasciurus hudsonicus*).

SQUIRREL, RED; PINE SQUIRREL; SPRUCE SQUIRREL; CHICKAREE *(Tamiasciurus hudsonicus)*

The Red Squirrel's range extends from most of Alaska, southward to include almost all of Canada except the Arctic regions, southern Alberta and Saskatchewan and Newfoundland. From British Columbia the range extends southward into the Rocky Mountains area, to Arizona and New Mexico. In the east the range includes New England, southward to Virginia and the Blue Ridge Mountain area, westward from New York and Pennsylvania north of the Ohio River to Minnesota and eastern North Dakota. Populations of Red Squirrel are also found in a small area of western Kentucky.

One of the smaller tree squirrels, and the smallest in its range, the Red Squirrel is from 10 to 15 inches (25 to 38 cm) long and weighs from about 5 to 9 ounces (142 to 255 g). The fur is a rusty red to grayish-red color and white to grayish-white below. The brightest red fur is on the sides. The bushy tail is similar in color to the back fur but has a broad black band outlined in white. During the summer months, the coat color dulls and a line of black fur separates the red upper fur from the white fur on the underside.

In coniferous forests, it feeds mainly on pine seeds, so much so that a forest floor littered with remains of pine cones is a sign of the presence of the Red Squirrel. Also included in the Red Squirrel's diet are acorns and other nuts, various tree seeds, birds' eggs, nestlings and fungi. In fall, the Red Squirrel buries large quantities of green pine cones and other foods.

Nests are made of shredded bark in hollow or fallen trees. Although it does not hibernate, it may stay in its nest during particularly bad winter weather. In the winter, it tunnels through the snow in a search for food.

Mating takes place in the winter, preceded by frenzied chases through trees. Litters of three to seven young are born in March or April, and there may be a second litter in August or September. The Red Squirrel is interesting to watch in its natural habitat, but too small to interest hunters. It sometimes gets into woodland vacation cabins, causing minor damage.

Similar species are the Gray Squirrel (*Sciurus carolinensis*), Fox Squirrel (*Sciurus niger*) and Douglas's Squirrel (*Tamiasciurus douglasii*).

SQUIRREL, SOUTHERN FLYING *(Glaucomys volans)*

The Southern Flying Squirrel is found in all of the United States east of the Mississippi River except Maine, northern Wisconsin, and Southern Florida. West of the Mississippi, it is found in the tier of states from Louisiana to Minnesota, and in the eastern parts of Texas, Oklahoma, Kansas, and Nebraska, and extreme eastern North and South Dakota.

The smallest of the tree squirrels, the Southern Flying Squirrel is from 8 to 10 inches long (20 to 25 cm) and weighs some 1¾ to 3½ ounces (50 to 100 g). The silky fur is grayish-brown above and white below. The tail is flattened and covered with gray-brown fur and there is a loose fold of skin between the front and hind legs.

The Southern Flying Squirrel does not actually fly, but it glides an impressive distance of 240 feet (73 m) or more with the help of the loose fold of skin between its legs. It leaps, spreads out its legs, stretching out the fold of skin so that it acts somewhat like a combination between an airfoil and a parachute. During the glide, it can change direction and rate of descent. It cannot, however, gain altitude. As it lands, it drops the tail and begins to retract the outstretched fold of skin, an action that makes an air-brake out of the fold. It lands lightly on a tree trunk with all four feet, immediately scurrying around to the other side of the tree.

Mating takes place in early spring, and litters of two to six young are born after a gestation period of 40 days. There is sometimes a second litter in the late summer. Sometimes kept as a pet, the Southern Flying Squirrel has no economic impact on human activity other than occasionally nesting in attics and causing minor damage.

The Northern Flying Squirrel (*Glaucomys sabrinus*), which is found in Canada and parts of the northern United States, is a bit larger and grayer in color than the Southern Flying Squirrel. The two species are similar in all other respects, including habits.

VOLE, MEADOW; FIELD MOUSE *(Microtus pennsylvanicus)*

The Meadow Vole is found in most of Alaska and Canada except the extreme northern regions, southward into the continental United States including all the northeastern states, the Atlantic Coast states southward to northeastern Georgia, all the states bordering the Great Lakes, West Virginia, eastern Kentucky and Tennessee, Iowa, eastern Nebraska, northern Missouri, North and South Dakota, Montana, most of Idaho and Wyoming, central Colorado, northwestern

New Mexico, north central Utah and northeast Washington.

The Meadow Vole is 5½ to 7¾ inches (14 to 20 cm) long and weighs about ¾ to 2½ ounces (21 to 71 g). The tail is from 1¼ to 2½ inches (3 to 6.3 cm) long. The body fur is variable, ranging from yellowish to reddish-brown peppered with black to blackish-brown in the upper regions. The fur on the underside is buffy, silvery or dark gray, and the feet are usually darker than the rest of the body. The overall body shape is somewhat ovoid.

Active throughout the day, the Meadow Vole's varied diet includes tubers, green plants, clover, plantain, seeds, grain, bark and insects. They eat almost constantly and are believed to consume almost their own weight daily. An alarmed vole will stamp its hind feet.

The Meadow Vole breeds throughout the year and is believed to have several litters annually of one to nine young.

The Meadow Vole and other voles can damage fruit trees by removing bark from roots and girdling the trunks. They have been seen to damage stored hay and grain. Voles are the major prey of a number of fur-bearing animals, including mink and ermine, and hawks and owls.

Similar species include the Mountain Vole (*Microtus montanus*), Prairie Vole (*Microtus ochrogaster*), and the Longtail Vole (*Microtus longicaudus*).

WALRUS
(Odobenus rosmarus)

The Walrus exists in two distinct populations or races, the Atlantic and the Pacific. The Atlantic race is found from Greenland south to Hudson Bay. Pacific Walruses are found in the Chukchi Sea off the coast of northeastern Siberia in the summer and the Bering Sea off southwest Alaska in the winter.

One of the larger of the sea-going carnivores, Walruses vary in size according to sex and race. The Pacific Walruses tend to be larger than the Atlantic Walruses and the males of both races are larger than the females. Atlantic males are from 8¼ to 11½ feet (2.5 to 3.5 m) long, and have an average weight of about 1600 pounds (726 kg). The Atlantic females range from 7½ to 9½ feet (2.2 to 2.9 m) in length and average some 250 pounds (568 kg) in weight. The larger Pacific males are as long as 12 feet (3.6 m) and can weigh as much as 3000 pounds (1362 kg). Female Pacific Walruses are only slightly larger than their Atlantic counterparts, and average about 1400 pounds (636 kg).

All walruses have hairless bodies yellowish to reddish brown in color that become pale after a long dive, when the blood goes from the surface capillaries to internal organs. The hair is limited to the muzzle 'beard' composed of about 400 one-inch (2.5 cm) bristles. The most distinctive features are the tusks, which are enlarged upper canine teeth. Tusks up to 40 inches (101 cm) long have been observed on Pacific bulls, while those of Atlantic bulls are usually around 16 inches (41 cm) long. Females also have tusks, but they are shorter, narrower and more curved in the middle than those of the bull.

Highly sociable animals, Walruses gather in herds of 2000 or more, bulls, females and juveniles separating by sex only during the breeding season. Herding Walruses defend each other against attacks by predators, and help those who are injured. The Walrus spends more time on land sunbathing and resting on beaches and ice than other sea-going carnivores. It also rests and sleeps at sea, hanging vertically in the water, using air sacs in the neck to stay afloat.

Although they appear to be ungainly, clumsy animals on shore, they are excellent graceful swimmers. Capable of easily swimming at speeds of 15 mph (24 kph), they can dive to depths of 300 feet (91 m) or more and remain submerged for as long as 30 minutes. Walruses often forage along sandy sea bottoms using their tusks and feeling out the sand with their sensitive muzzle bristles for clams and other mollusks and crustaceans which are

Above: **This Walrus bull (center) has a broken tusk.**

their major food. In the autumn months they may feed in shallow, ice-covered waters, keeping air holes in the ice open by butting through with their heads. Walruses eat clams much like people at a clam bar, sucking the meat off the shell with their dome-shaped mouths. An adult Walrus may eat 100 pounds (45 kg) or more of food a day, but in good weather it can go for a week or so without eating.

The tusks are used for defense, mating fights, hunting and other purposes such as providing a pulling anchor for sliding on ice and snow. Walruses occasionally take a seal by stabbing it with a tusk. The major predators, polar bears and man, soon learn to respect the deadly potential of the Walrus's tusks. The size of tusks establishes a bull's dominance.

Mating and courtship behavior begin at sea in April and May. Bulls may engage in courtship fights during which the tusks may be broken. At mating time, males become very vocal, emitting pealing bell-like sounds in addition to the usual bellows and grunts. A single pup, weighing from 100 to 150 pounds (45 to 68 kg) is born about a year later. Female Walruses mate every other year.

WHALE, GRAY
(Eschrichtius gibbosus)

A Pacific Ocean species, the Gray Whale migrates from the Arctic Ocean along the west coast of North America to the Gulf of California.

A medium-sized baleen (toothless) whale, the Gray Whale measures up to 45 feet (14 m) in length and weighs

***Above:* A Gray Whale spouts and dives off the California coast.**

from 34,000 to 74,000 pounds (15,000 to 33,000 kg). The body color is a mottled splotchy grayish-black. With a more slender looking body than most other baleen whales, the Gray Whale's spouts are low on the body and it has distinctive longitudinal folds on the throat. Instead of teeth, there are strips of baleen or whalebone hanging from the roof of the mouth. The baleen strips, fringed at the inner side, serve to strain the tiny marine organisms that are the whale's food out of the water. The organisms stick to the baleen and the whale licks them off with its massive tongue and swallows them.

The Gray Whale goes on the longest yearly migration of any mammal. Starting in late December, Gray Whales leave the Bering Sea and start to move southward to the coast of Baja, California, where they deliver their young and mate. They move back to the Bering Sea in March and April, covering a total of 16,000 miles (25,600 km).

The migration of the Gray Whale has become a tourist attraction. Every year, boats from the West Coast carry tourists to view the whales on their annual migration. Tourists also travel to the shallow water off Baja, California, to observe the Gray Whales' courting behavior that includes leaping out of the water.

The Gray Whale was once one of the more widely hunted whales. From time to time, kills were restricted or banned by the International Whaling Commission, but these were often ignored by the major whaling nations. The total population was estimated to be between 100,000 to 200,000 in the mid 1970s.

Similar species include the Right Whale (*Eubalaena australis*), the Finback Whale (*Balaenoptera physalus*) and the Humpback Whale (*Megaptera novaeangliae*).

WHALE, KILLER; SEA WOLF *(Orcinus orca)*

The Killer Whale (also called Orca, Grampus and Sea Wolf) is found in warm and cold oceans, close to the coast. It frequents bays, inlets and coves, particularly on the Pacific northwest coast of North America.

The largest of the dolphins, the Killer Whale can be as long as 30 feet (9.1 m). Females are several feet shorter than the males. Killer males have a distinctive pointed dorsal fin that can be as high as 6 feet (1.8 m) in the male and is usually about 3 feet (.9 m) high in an adult female. The fin points straight up on the male, and is somewhat hooked on the female. Like all the toothed whales, the Killer Whale has a single blowhole rather than the two blowholes seen in the baleen whales. The Killer Whale is dark black on the upper body regions and a true white on the sides and underside. A white spot is present under each eye. The snout is rounded, and an open mouth reveals rows of 10 to 15 sharp teeth, on each side of the upper and lower jaws.

The Killer Whale is the only whale known to regularly eat sea mammals including seals, sea lions and other whales, particularly baleen whales. In fact, whales have no other natural enemies besides man. It may also take water birds such as penguins. Capable of speeds of 24 mph (38 kph) it can outpace most prey. It also eats fish such as tuna. Killer Whales often hunt cooperatively, herding fish and other prey into blind inlets. They travel in family groups, called pods, of ten to fifty individuals, and often play and cavort with each other. They are sometimes seen to hold their bodies upright in the water, with the head above the surface in an attitude referred to as 'spying.' Often done by several killer whales in a neat row, spying is believed to be a way of scanning the water for prey. In general, however, whales depend more on their keen sense of hearing than on their sight.

Believed to mate for life, Killer Whales frequently engage in sexual play. After a gestation period of at least 12 months, the female delivers a single young, born under water. They call to each other with distinctive wailing cries which are believed to have specific meanings. Observers of these whales soon learn to recognize individuals by their vocalizations.

Although seafarers' folklore is filled with tales of Killer Whales attacking and eating humans, there are very few documented cases of such attacks, and these are primarily attacks by wounded animals. While the Killer Whale can be quite ferocious when attacking prey in the wild, it is docile and friendly in captivity. Intelligent, trainable animals, their use in seaquarium shows has increased in recent years. It was never an important species in the whaling industry, except in Japan and Norway, so the Killer Whale population is believed to be holding steady.

WOLF; GRAY WOLF; TIMBER WOLF
(Canis lupus)

The range of the Gray Wolf once extended over almost all of North America. Today, the Gray Wolf is found in Alaska, almost all of Canada, including the Arctic Ocean Islands, with the exception of Quebec east of the St Lawrence River, the Maritime Provinces and Newfoundland. It is also found in coastal Greenland. In the United States, the Gray Wolf is limited to Michigan's Upper Peninsula, extreme northern Minnesota and Wisconsin, and scattered populations in Wyoming, Colorado, Utah, Montana, Idaho, Washington, Texas, Arizona and New Mexico. A well-studied population exists in Isle Royal National Park in Lake Superior.

The Gray Wolf is 2 to 3 feet (.6 to .9 m) high at the shoulder, 3.3 to 6¾ feet (1 to 2 m) long, and weighs about 60 to 130 pounds (27 to 59 kg). Males are larger than the females. Fur varies from almost white to black, but a grizzled gray is the usual color.

The Gray Wolf lives and hunts in tundra and forest in groups or packs of two to 15 individuals. Usually composed of family groups, the packs are generally led by the strongest male.

Below: **The Gray or Timber Wolf will often live and hunt in a pack.**

Packs hunt in territories of approximately 100 to 250 square miles (259-648 sq km) that generally overlap with those of other packs. Usually hunting at night, the pack uses cooperative tactics to bring down prey which are usually large herbivores such as deer, moose and caribou. However, they will also eat a variety of other foods including smaller mammals, birds, fish, insects and even berries and other fruit. Since healthy large mammals such as deer can usually outrun Wolves, they seldom run after prey for long distances, preferring to surround the victim and drive it back to other wolves waiting in ambush. Wolves can usually capture a healthy adult moose, deer or other large animal if the prey becomes mired in deep crusted snow. Wolves will generally be supported by the crust while the larger, heavier animals may fall through and flounder. Wolves usually take young, slower-moving animals or older ones that are sick or injured.

One of the most distinctive features of the Gray Wolf is its howling. Often heard at night, the howling may be a preliminary to the hunt. As soon as one Wolf starts to howl, the others in the pack join in. Often they gather in a circle, and stragglers bound in, apparently eager to join the howl. They have been observed to change the pitch of the howls, in what could be interpreted as an attempt to harmonize. In addition to howling, Wolves bark, yap, whine and growl. Wolves separated from the pack signal their location with the 'lonesome call,' a short howl that rises in pitch and then fades.

Mating takes place in February and March. In some packs, only the dominant male mates. Litters of five to 14 pups are born in April and June. The black-furred pups come out of the den to play at about age one month. All members of the pack take part in caring for the young. When pack members returning from the hunt are nipped on the snout by the pups, the hunters regurgitate undigested meat for them.

WOLVERINE; GLUTTON
(Gulo luscus)

The Wolverine's range includes almost all of Alaska, the northern half of Canada from Yukon to Labrador, almost all of British Columbia and scattered populations in Colorado, California and possibly other Rocky Mountain states.

Looking somewhat like a small, chunky bear, the Wolverine (a member of the weasel family) is from 2.6 to 3.6 feet (.8 to 1.1 m) long and weighs from about 18 to 45 pounds (8 to 20 kg). Males are larger than females. The fur is dark brown and broad yellowish bands run from the shoulders to the rump and meet at the base of the tail. There are light patches of fur over the ears. The feet are rather large for the size of its body.

Although it is a determined, ferocious and skilled killer, it prefers carrion to fresh kills. However, when carrion is not available it will go after any animal it can find and kill, including moose, elk, caribou, deer, beaver, porcupines and rodents, birds and eggs. It also eats plant foods such as roots and berries. It often eats trapped animals and other animals' food caches. The name 'glutton' stems from a voracious appetite, which is probably an adaptation for survival over periods of food scarcity. It carelessly caches

***Above:* The Woodchuck or Groundhog often falls victim to hunters and cars as well as predators like the Red Fox.**

food it cannot immediately eat, often marking it with a foul-smelling anal gland musk, urine and feces.

The Wolverine covers a wide area of 1600 square miles (4144 sq km) or more in its search for food. Typically, a hunting territory of that size serves a male and two or three females, and overlaps with those of other Wolverines. The Wolverine moves in a slow lope, but can cover great distances in intense three to four hour periods of activity, alternating with rest periods. It is a good swimmer and an efficient tree climber, often pouncing on prey from trees. It builds dens in protected places such as crevices and under uprooted trees, but does not hibernate.

Wolverines have a long mating season, extending from April to September. This long season increases the possibility of widely dispersed Wolverines meeting at a time when the female is receptive. Litters of two to three are born from February to April. Females are believed to have a litter every two or three years. The young remain with the mother for at least two years before leaving to establish their own territories.

Wolverines have been known to raid cabins and campers' tents, eat their fill and render what they don't eat useless by marking it with musk, urine and feces. Wolverine fur has oils in it that make it frost resistant, a quality that encourages its use as a lining for parkas.

A lone Wolverine can successfully drive off large predators, including bears and mountain lions, from its kills.

WOODCHUCK; GROUNDHOG; MARMOT
(Marmota monax)

The range of the Woodchuck extends from east central Alaska southeastward to include most of southern Canada. The range extends southward from Canada to include the northeastern states of the United States including the Middle Atlantic states, all of the states bordering the Great Lakes, Iowa, Missouri, Arkansas, Kentucky, West Virginia, Virginia, most of Tennessee, western North Carolina and northern Georgia and Alabama. It is also found in northeastern North Dakota and the extreme northern 'leg' of Idaho.

One of the larger rodents, the Woodchuck is from 1⅓ to 2½ feet (.4 to .75 m) long, and weighs from about 4½ to 14 pounds (2 to 6.3 kg). The fur varies from grizzled brown to reddish to blackish. The fur on the belly is often paler than that of the upper parts of the body. The Woodchuck has a stocky body and short, powerful legs well adapted for digging.

The diurnal Woodchuck is particularly active on sunny days, usually in the morning and later afternoon. Feeding primarily on vegetation, including grasses, clover, plantain and alfalfa, it will also feed on garden and crop plants

An efficient digger, the Woodchuck digs burrows as much as 5 feet (1.5 m) deep and 30 feet (9.1 m) long. The burrow may have several tunnels, each of which may end in an enlarged nesting chamber. Woodchuck nesting chambers are frequently used by other mammals, including raccoons, skunks, rabbits, foxes and others who may use the burrow as the Woodchuck made it,

***Above:* A Bushy-tailed Woodrat or Pack Rat.**

or enlarge it to suit their own needs.

It is a good swimmer and will nimbly climb a tree to escape a predator, but it prefers to stay near its den. It has a variety of vocalizations including a shrill whistle, hisses, growls and teeth chattering. The latter is a sign of anger.

A true hibernator, the Woodchuck begins to put on a layer of fat in late summer and early fall. As the winter approaches, it retires to its burrow, curls up in a tight ball on a mat of grass, and falls into hibernation. During hibernation, the body temperature falls from the normal range of 97°F (36°C) to 40°F (4.4°C), and the heartbeat drops from about 100 to 4 beats per minute. The emergence of the Woodchuck from its hibernation burrow in early spring is the basis of the 'Groundhog Day' observance of February 2. The actual day of emergence varies with conditions and latitude; in the northern parts of the range, it usually emerges in March.

The mating season starts as soon as the Woodchuck emerges from hibernation. A male seeks out a receptive female in its burrow, remaining with her a short time. Litters of four to five young are born in April or early May. Although they do not open their eyes and crawl until they are one month old, they are ready to go out on their own when they are only two months old.

Woodchucks are among the more widely hunted of 'varmints,' and large numbers are killed by automobiles. Woodchucks can damage crops and gardens; however, their digging and excrement serve to aerate and fertilize the soil.

Similar species include other marmots such as the Yellow-bellied Marmot (*Marmota flaviventris*).

WOODRAT, BUSHY-TAILED; PACK RAT; MOUNTAIN PACK RAT
(Neotoma cinerea)

The Bushy-tailed Woodrat is found from British Columbia, extreme southwest Yukon and extreme western and southern Alberta, southward into the northwestern continental United States from Washington south to northern California, eastward to include most of Nevada, all of Utah, western Colorado, extreme northern Arizona and New Mexico, and northward to include Wyoming, Montana, Idaho and western North and South Dakota.

The Bushy-tailed Woodrat is 11 to 18 inches (28-46 cm) long and weighs from 5 to 16 ounces (142-454 g). The bushy, but somewhat flattened tail of 4½ to 9 inches (11-23 cm) long can be as much as half the total body length. Variable in color, most are pale-grayish to blackish above and whitish below. Some have brown fur sprinkled with black.

Like all woodrats, the Bushy-tailed Woodrat stores and 'trades' a variety of objects, which it stores in houses built of sticks in concealed spots in rocky crevices. It may also build its house fairly high in trees in its rocky coniferous forest habitat, but the tree house nests are used only for storing dried vegetation. With a strong preference for shiny objects, it will drop whatever it is carrying and 'trade' it for something shiny such as a coin or other piece of metal.

The Bushy-tailed Woodrat eats a variety of vegetation including leaves, stems, nuts, seeds and mushrooms, and may occasionally eat the flesh of the small mammals. Excellent climbers, they use their bushy tails for balance. They thump their hind feet on the ground, particularly when alarmed, but will also thump at other times.

Not found in areas of dense human population the Bushy-tailed Woodrat has no impact on human activities.

Similar species include the White-tailed Woodrat and the Desert Woodrat (*Neotoma lepida*).

BIBLIOGRAPHY

Fisher, James, Noel Simon, and Jack Vincent. *Wildlife in Danger.* N.Y. Viking Press, 1969

Grzimek, Bernhard, ed. *Grzimek's Animal Life Encyclopedia*, 13v. N.Y. Van Nostrand-Reinhold, 1972-75.

Handbooks of American Natural History. Ithaca, N.Y. Comstock, 1942 –.
especially the following titles:
Comstock, A. *Handbook of Nature Study.* 1939.
Hamilton, W. J. *The Mammals of Eastern United States.* 1943.

Jarman, Cathy. *Atlas of Animal Migration.* John Day, 1972.

Larousse Encyclopedia of Animal Life. N.Y. McGraw-Hill, 1967.

Leftwick, A. W. *Dictionary of Zoology.* 3d ed. London, Constable, 1973.

Rand McNally Atlas of World Wildlife. Rand McNally, 1973.

The Naturalists' Directory International. South Orange, N.J., Naturalists' Directory, 1876 –. Irregular. (43rd. ed. 1977)

Palmer, E. Laurence. *Fieldbook of Natural History.* 2d ed. Revised by H. Seymour Fowler. N.Y.: McGraw-Hill, 1975.

Peterson, Roger Tory. *The Peterson Field Guide Series.* Sponsored by the National Audubon Society and the National Wildlife Federation. Boston. Houghton, 1947 –
especially the following titles:
Field Guide to the Mammals: Field Marks of all North American Species Found North of the Mexican Boundary. 3rd ed. 1976.
Field Guide to Animal Tracks. 2nd ed. 1975.

Thompson, John W. *Index to Illustrations of the Natural World: Where to Find Pictures of the Living Things of North America.* Syracuse, N.Y.: Gaylord, 1977.

Desmond, Morris. *The Mammals: A Guide to the Living Species.* N.Y.: Harper, 1965.

Hall, Eugene Raymond and Keith R. Kelson. *Mammals of North America.* 2v. New York, Ronald, 1959.

Walker, Ernest P. *Mammals of the World.* 3rd ed. Baltimore Johns Hopkins University Press, 1975. 3v

LIST OF ENDANGERED SPECIES* (NORTH AMERICA)

*US Dept of the Interior January 1982

MAMMALS

Bat, gray (Central and Southeastern US)
Bat, Indiana (East and Midwestern US)
Bat, Ozark big-eared (MO, OK, AR)
Bat, Virginia big-eared (KY, WV, VA, IN, IL, OH)
Bison, wood (Can., Northwest, USA)
Cougar, eastern (Eastern US)
Deer, Columbian white-tailed (OR, WA)
Deer, key (FL)
Ferret, black-footed (Western US)
Fox, Northern swift (USA, Can.)
Fox, San Joaquin kit (CA)
Jaguar (TX, NM, AZ)
Jaguarundi (TX, AX)
Manatee, West Indian (Florida) (Southeastern US)
Margay (NM, AZ)
Mouse, salt marsh harvest (CA)
Ocelot (USA (SW))
Panther, Florida (LA and AR east to SC and FL)
Prairie Dog, Utah (UT)
Pronghorn, Sonoran (AZ)
Rat, Morro Bay kangaroo (CA)
Seal, Carribbean monk (Gulf of Mexico)
Squirrel, Delmarva Peninsula fox (MD, VA, DE)
Whale, blue (Oceanic)
Whale, bowhead (Oceanic)
Whale, finback (Oceanic)
Whale, gray (Oceanic)
Whale, humpback (Oceanic)
Whale, right (Oceanic)
Whale, Sei (Oceanic)
Whale, sperm (Oceanic)
Wolf, gray (48 conterminous States, other than MN)
Wolf, red (Southeast US west to TX)

PICTURE CREDITS

N P Baldachino, USFWS 58 (bottom).
California Academy of Sciences (CAS) 54 (top).
Colorado Department of Natural Resources, Division of Wildlife 63.
Florida Department of Commerce 25.
Dick Frear, NPS 12, 28 (top), 50, 51 (bottom).
R. V. Hansen, USFWS 47.
George Harrison, USFWS 32 (top).
Jerry L Hont 39 (top).
Lloyd G Ingalls 11, 22, 38, 39 (bottom), 42 (both), 49, 57, 60 (all).
Dave Olson, USFWS, 15 (left both).
Danny On, Glacier Natural History Association 2-3, 13, 24, 34.
Edwin Rothfuss 35 (top).
Leonard Lee Rue 36.
V B Scheffer 10 (both).
Richard Simms, Tennessee Wildlife Resources Agency 15 (right), 16 (left), 27 (bottom), 35 (bottom), 41 (bottom), 56, 62.
Jim Jones, USFWS 9 (bottom).
William B. Keller, NPS 37 (bottom).
Jim Leupold, USFWS 19.
Wayne Lynch, PC 8, 9 (top), 20, 28 (bottom), 32 (bottom), 45 (left), 55, back cover.
Keith Mouhouse, USFWS 27 (top).
National Park Service 16 (right), 17 (top), 26, 43, 44, 59, 61, 63.
Doug Storer 4-5.
US Department of Agriculture 48.
US Fish and Wildlife Service (USFWS) 41 (top).
Utah Division of Wildlife Resources 21, [illegible]1, 33, 45 (right).
M Wick 51 (top).